Radiology of Musculoskeletal Stress Injury

Other Books by Theodore E. Keats

An Atlas of Normal Developmental Roentgen Anatomy, Second Edition, with Thomas A. Smith
Emergency Radiology, Second Edition
Atlas of Roentgenographic Measurement, Fifth Edition, with Lee B. Lusted
Atlas of Normal Roentgen Variants That May Simulate Disease, Fourth Edition

Radiology of Musculoskeletal Stress Injury

THEODORE E. KEATS, M.D.
Professor and Chairman
Department of Radiology
University of Virginia School of Medicine
Charlottesville, Virginia

YEAR BOOK MEDICAL PUBLISHERS, INC.
CHICAGO • LONDON • BOCA RATON • LITTLETON, MASS.

2   3   4   5   6   7   8   9   0   Y   R   94   93   92   91   90

Library of Congress Cataloging-in-Publication Data

Keats, Theodore E. (Theodore Eliot), 1924-
    Radiology of musculoskeletal stress injury / Theodore E. Keats.
        p.   cm.
    Includes bibliographical references.
    ISBN 0-8151-5002-4
    1. Musculoskeletal system—Wounds and injuries—Diagnosis.
2. Overuse injuries—Diagnosis.   3. Musculoskeletal system—
Radiography.   4. Diagnosis, Radioscopic.   I. Title.
    [DNLM: 1. Bone and Bones—injuries.   2. Diagnostic Imaging.
3. Fractures.   4. Muscles—injuries.   5. Repetition Strain Injury.
WE 180 K25r]
RD734.5.R33K43   1990
617.4'710757—dc20   89-22456
DNLM/DLC   CIP
for Library of Congress

Sponsoring Editor: James D. Ryan
Assistant Director, Manuscript Services: Frances M. Perveiler
Production Project Coordinator: Gayle Paprocki
Proofroom Supervisor: Barbara M. Kelly

To my wife, Patt,
who helps me with stress of all kinds.

Preface

*There is often a very fine line between
physical fitness and madness.*

T.E. Keats

Musculoskeletal stress injuries were once thought to be largely a disease of soldiers or athletes and, in these settings, were relatively easy to diagnose. With the new emphasis on physical fitness, these injuries are being seen with increasing frequency in children and adults and in locations that are not considered usual sites of stress injury. In addition to the multitude of stress-induced injuries of the normal skeleton, there are a variety of insufficiency fractures of the osteoporotic skeleton which have been recognized only relatively recently and which are often overlooked or misdiagnosed.

It is the intent of this work to bring together in a single compendium the current state of knowledge concerning the radiologic manifestations of the entire spectrum of abnormalities that result from muscular stress on the skeleton. For this writing I have endeavored to cull from an enormous literature those sources which are most timely and authoritative. Since the diagnosis of these lesions is largely radiologic, accuracy in diagnosis is crucial, particularly since the radiologic manifestations may closely resemble those of infection or neoplasm. I have seen such errors of commission lead to many unnecessary investigations, biopsy, surgical excision, and even amputation.

The reader will note that a small number of radiographic reproductions are substandard in quality. These represent unusual entities taken from the older literature. Unfortunately, the original films or prints of these cases are no longer available, and the images presented here are photographs of the illustrations as they appeared in the original journal. Although lacking in fine detail, they still document the particular injury.

I would like to express my appreciation to the many contributors of illustrative material which has permitted me to expand my presentation beyond my own case material. I wish also to thank my editorial assistant, Carol Chowdhry, and my secretary, Patricia West, for their expert contributions. Thanks are also due to Janeth Bibb, Edie Pollard, Donna Marshall, Charlene Bolling, and Leigh Sarkozi for superb radiography.

THEODORE E. KEATS, M.D.

Contents

Introduction

Injuries to the musculoskeletal system occur in normal bones as well as in those affected by osteoporosis or osteomalacia. The location of these injuries depends on the age of the patient as well as the nature of the inciting activity. The common denominators of all the injuries that fall into the spectrum of musculoskeletal stress are absence of external trauma and a history of muscular effort of some kind.

In 1855, Breithaupt,[1] a Prussian army surgeon, described a painful syndrome of the foot that developed in soldiers after long marches. He ascribed it to an inflammation of the synovial sheaths. In 1887, Pauzet,[2] an army physician, also described this syndrome in young infantrymen and noted marked periosteal proliferations of the metatarsals, an observation based on clinical examination only. He also noted that exostoses developed on the second, third, and fourth metatarsals, most commonly on the second. He related these periosteal changes to the design of the military boot. Stechow,[3] a member of the Prussian Guard, first described the radiographic appearance of stress fracture in 1897. He described 36 cases and identified a fracture of the metatarsal as the underlying pathologic lesion. In 1905, Blecher[4] reported the first case of stress fracture of the femoral neck. In 1921, Deutschlander[5] reported a comprehensive study of stress lesions and pointed out their occurrence in children and adults over the age of 50. In the German medical literature, stress fracture of the metatarsal is called Deutschlander's disease. Asal[6] published the first large series of stress fractures in 1936, describing 590 stress fractures among German troops.

Since these early reports, the literature on the subject has grown voluminously, in part stimulated by World War II. These data demonstrated that all patients of all ages are affected and that many different parts of the skeleton may be involved. To date, the two definitive books on the subject of stress fractures are those of Morris and Blickenstaff in 1967[7] and Devas in 1975.[8] The concept of stress-induced alteration in bone and muscle has been expanded beyond the stress fracture and forms a sizable portion of this work.

ETIOLOGY OF STRESS-RELATED INJURY

Early concepts of stress fracture as a fatigue injury suggest that the problem is present in weight-bearing bones only. It is now well recognized that these injuries occur throughout the skeleton and are related to overuse rather than weight bearing alone. These stress lesions appear to be reflections of a similar process regardless of location, presentation, or age of the patient.

As the histologic studies of Johnson et al.,[9] indicate, stress fractures result from a succession of connected stresses that create an imbalance in the physiologic process of bony remodeling. These mechanisms have been recently detailed in an excellent review by Markey,[10] and the salient features of his review are restated here.

The concepts of stress and strain can be related to engineering and also to bone. The physical properties of cadaver bone can be tested mechanically in the laboratory to determine the stress failure point. However, in life these stresses are compounded by muscular arrangements that result in additional stresses. There are axial, rotary, and bending movements applied to bone from each end, but there are also those directed to the middle segment, through muscle attachment. These primary and residual forces stress the bone, which must redesign itself to resist.

The piezoelectric (electric polarity due to pressure, especially in a crystalline substance) effect of bone crystal response to stress has been implicated as a factor in the reaction of bone to stress. Relative electropolarity results from tension forces and is supportive of osteoclastic activity and resultant bone resorption. On the other hand, compressive forces develop an electromagnetic field that stimulates osteoblastic new bone formation. Most cortical stresses are due to tension, since torsion produces tension circumferentially and all bending movements produce tension and compression. Electromagnetic studies have shown that such tension forces result in microfracture and debonding at the cement lines and interlamellar cement bonds. These two forces, which resorb and replace bone, are con-

TABLE 1.

Distribution of Stress Fractures*

Bone Involved	Total No. of Cases	Percentage
Metatarsal	88	35.2
Calcaneus	70	28.0
Tibia	60	24.0
Ribs	14	5.6
Femur	8	3.2
Fibula	8	3.2
Spine	1	0.4
Pubic ramus	1	0.4

*From Wilson ES Jr, Katz FN: Stress fractures: An analysis of 250 consecutive cases. *Radiology* 1969; 92:481. Reproduced by permission.

stantly interacting, Excessive tension forces may exceed compressive ones and lead eventually to fracture.

Muscle attachments to bone also contribute to cyclic stress. Regular cyclic stress by muscular activity will induce remodeling, and overuse will result in cortical thinning and/or osteoporosis, with the development of a stress fracture if the cyclic stress is continued.

In some bones, such as the metatarsal, the tarsal navicular, and the femoral neck, muscular attachments play little role and weight bearing alone produces the repeated stress.

Another popular theory holds that muscle fatigue is a contributing factor in that muscles fatigue during repetitive exercise and are less effective in absorbing impact. More force is therefore transmitted directly to bone. If this concept is indeed true, stress fractures of bone may represent a failure of muscle function first, and bone failure is a secondary event.

ANATOMIC DISTRIBUTION

Much of the early data on the incidence and anatomic distribution of stress fractures came from military experience, and these series are the largest available in the literature. Wilson and Katz[11] surveyed 250 cases encountered during a 10-month period at a large military training center. Their results, given in Table 1, show the metatarsal to be the most frequent site of injury, followed closely by the calcaneus and tibia. In a larger series of Marine recruits, Greaney et al.,[12] surveying initially by scintigraphy, showed the tibia to be most commonly involved, followed by the posterior calcaneal tuberosity (Table 2).

It is of interest that if one studies the anatomic distribution of stress fractures in athletes, the location of injury is much the same as that seen in soldiers. Matheson et al.,[13] in a study of 320 athletes by bone scanning, showed that the most common bone injured was the tibia (49.1%), followed by the tarsals (25.3%), metatarsals (8.8%), femur (7.2%), fibula (6.6%), pelvis (1.6%), sesamoids (0.9%), and

spine (0.6%). Running was the most common sport at the time of injury.

McBryde[14] compared the incidence of stress injury in athletes and that seen in military recruits. Table 3 summarizes his results. The distribution in athletes is not dissimilar from the military experience except for the incidence of upper extremity lesions. It has been proposed that this change in peak incidence from the metatarsals to the tibia as the primary site of stress injury is due to emphasis on running rather than marching, a change in footwear, and a more fit group of recruits. McBryde[14] suggests that the incidence difference in stress fractures in men vs. women and whites vs. blacks and other ethnic backgrounds is on the basis of duration and repetition of activity rather than biologic differences.

As anticipated, there is a correlation between sport or athletic activity and the osseous location of the injury. Orava et al.[15] present these kinds of data (Table 4 and Fig 1). The largest incidence occurs in track and field events, where the tibia is again the most common site of stress injury. A listing of sports and their associated injuries is given in Table 5.

The spectrum of injury resulting from musculoskeletal stress is limited only by the ingenuity of the exerciser, as our discussion of individual sites of injury will show.

CLINICAL ASPECTS

The characteristic clinical presentation of all stress injuries is pain at the site of injury that is relieved completely or partly by rest. The pain may be abrupt in onset, as with an avulsive injury, or insidious, extending over a 2- to 3-week period, gradually increasing in severity with continued use. Although pain relief with rest is the usual sequence, relief may not be complete in that there may be residual aching, even with rest.

The painful area is usually localized and tender to palpation. Stress injuries of the hip may exhibit pain referred to the knee or leg. The fact that the pain and tenderness are located over the periosteum rather than the ligaments strongly suggests stress injury rather than ligamentous sprain.[7]

Swelling of the involved area is often present and is also localized. However, in some cases of stress fracture of the tibia or metatarsal, a large area of swelling may be evident. Occasionally, there may be swelling and effusion in an adjacent joint. Swelling and erythema together may be present, particularly in children and the very old.[8] A palpable mass due to formation of callus may be detected in older injuries.

There is usually no fever, leukocytosis, or elevation of the erythrocyte sedimentation rate. When present, such findings are usually coincidental.

In dealing with athletes or soldiers, one's index of suspicion for a stress-related injury is very high. However,

TABLE 2.

Anatomic and Intraosseous Distribution of Initial Scintigraphic Abnormalities*†

Site	Total No. of Patients	Total No. of Abnormal Sites	Total Ab. Radiographs	3+ Sites/Ab. Radiographs	2+ Sites/Ab. Radiographs	1+ Sites/Ab. Radiographs
Leg						
Femur (trabecular)	36	48	3	4/0	15/2	29/1
Femur (cortical)	27	48	9	3/2	15/3	30/4
Patella	8	12	0	0	6/0	6/0
Fibula	5	8	2	3/0	4/2	1/0
Tibia (trabecular)	120	192	34	50/17	65/10	77/7
Tibia (cortical)	103	124	59	54/43	34/10	36/6
Total	**299**	**432**	**107**	**114/62**	**139/27**	**179/18**
Foot						
Talus	33	46	10	18/4	21/4	7/2
Calcaneus	95	172	83	96/53	43/17	33/13
Navicular	20	28	12	9/3	17/7	2/2
Cuboid	11	13	2	7/2	5/0	1/0
Cuneiforms	6	6	2	5/2	1/0	0
MT (trabecular)	62	133	16	58/8	65/6	10/2
MT (cortical)	8	9	6	3/3	0	6/3
Total	**235**	**407**	**131**	**196/75**	**152/34**	**59/22**
Trabecular						
Leg	164	252	37	54/17	86/12	112/8
Foot	227	398	125	193/72	152/34	53/19
Total	**391**	**650**	**162**	**247/89**	**238/46**	**165/27**
Cortical						
Leg	135	180	70	60/45	53/15	67/10
Foot	8	9	6	3/3	0	6/3
Total	**143**	**189**	**76**	**63/48**	**53/15**	**73/13**
Total	...	**839**	**238**	**310/137**	**291/61**	**238/40**

*From Greaney RB, Gerber FH, Laughlin RL, et al: Distribution and natural history of stress fractures in U.S. Marine recruits. *Radiology* 1983; 146:339. Reproduced by permission.
†*Ab.* = abnormal; *MT* = metatarsal.

with an ordinary patient, a more careful history is essential for correct diagnosis. The key is the elucidation of some activity or change of activity out of the ordinary for that individual. This change may be in kind or degree. This is particularly important in dealing with the middle-aged woman or elderly patient in whom the most innocent change in activity may lead to an insufficiency fracture.

Even in the seasoned athlete, an alteration in the pattern of exercise may coincide with the onset of symptoms. Longer training periods, holidays, longer runs, different athletic gear, changes in road or track surfaces—all may be implicated.

Inquiry concerning the patient's occupation may be helpful, particularly in the individual whose source of stress is not obvious or whose bones or joints show evidence of overuse. The cornerstone of early diagnosis of stress injury is to suspect it.

TABLE 3.

Approximate Incidence of Stress Fractures*

Anatomic Location	Athletes, %	Military, %
Fibula	25	2
Metatarsal	20	40
Tibia	20	20
Os calcis	15	30
Femur	10	3
Talus	...	...
Ribs	...	...
Spine	6	5
Pubis	...	...
Upper extremity	3	...
Other	1	...

*From McBryde AM: Stress fractures in athletes. *J Sports Med* 1975; 3:212–217. Reproduced by permission.

TABLE 4.

Relationship Between Sporting Activity and Stress Fracture Site*

Sports	Tibia	Metatarsal Bones	Fibula	Femoral Neck	Femoral Shaft	Pubic Arch	Metacarpal Bones	Sesamoid Bones	Humeral Shaft	Ulna	Tarsal Navicular	Toe Phalanx	Vertebral Arch	Total
Track and field	50	19	10	4	1	...	...	...	1	...	...	...	...	**85**
Jogging	7	5	5	...	1	2	...	...	...	...	...	1	...	**21**
Skiing	6	1	1	1	2	...	...	1	...	...	...	...	...	**12**
Orienteering	7	...	2	...	...	...	...	...	...	...	...	...	...	**9**
Ball games	4	1	1	...	...	...	...	...	...	1	1	...	...	**8**
Power events	...	...	...	...	...	...	2	...	...	...	...	...	1	**3**
Gymnastics	2	...	...	...	...	...	...	...	...	...	...	...	...	**2**
Cycling	...	...	1	...	...	...	...	1	...	...	...	...	...	**2**
Total	**76**	**26**	**20**	**5**	**4**	**2**	**2**	**2**	**1**	**1**	**1**	**1**	**1**	142

*From Orava S, Puranen J, Ala-Ketola L: Stress fractures caused by physical exercise. *Acta Orthop Scand* 1978; 49:19. Reproduced by permission.

RADIOLOGY

The correct diagnosis of stress-induced injury is based largely on radiologic findings. Correct diagnosis is critical for two reasons. First, the periosteal reaction that accompanies some of these injuries, particularly the stress fracture, may closely simulate the changes seen in malignant neoplasia, particularly osteosarcoma. Mistaking a stress injury for malignant neoplasm may in turn lead to biopsy and the second mistake, namely misinterpretation of repairing bone as osteosarcoma by an inexperienced pathologist.

The classic radiologic signs of stress fracture do not become evident until 1 to 2 weeks after the onset of symptoms. Before this interval, earlier confirmation can be obtained with nuclear scintigraphy. Bone scans will show a "hot spot" in the injured area within 1 to 4 days after onset of symptoms. Scintigraphy will be discussed in a subsequent section.

Various classifications of the radiologic signs of stress fractures have been offered.[16, 17] However, it appears that the sequence of events differs according to the bony site. Wilson and Katz[11] have offered a categorization of the initial roentgenographic features as falling into one of four basic patterns. These are:

Type I: Fracture line visible with no evidence of endosteal callus or periosteal reaction.

Type II: Focal sclerosis of bone as the manifestation of trabecular condensation within cancellous bone and formation of endosteal callus.

Type III: Periosteal reaction and external callus.

Type IV: Mixed combinations of the above.

Early stress fracture of the metatarsals and the neck of the femur are good examples of the type I stress fracture (Fig 2). The fracture may be linear or comminuted, with or without displacement of the fragments.

Typical examples of the type II classification include stress fractures of the calcaneus and the medial tibial plateau (Fig 3). The type III pattern of periosteal reaction and external callus is the pattern most likely to be confused with a malignant bone tumor in a long bone, although in the metatarsal march fracture this mistake is less likely. The presence of a radiolucent line within otherwise normal cortical bone will usually resolve the problem (Fig 4).

Healing of the stress fracture is characterized by callus formation progressing slowly to form a fusiform area of maturing callus, disappearance of the fracture line, and formation of areas of bone condensation. The variable patterns of healing will be in keeping with the type of injury as described above.

The differential diagnosis of stress injury is limited. Benign tumors such as osteoid osteoma, Ewing's sarcoma, and, particularly, osteogenic sarcoma can create problems in diagnosis. Generalized conditions such as infant abuse, infantile cortical hyperostosis, and vitamin deficiencies are usually multifocal and are not likely to be sources of confusion. In children, low-grade fever may accompany a stress fracture, and osteomyelitis might be considered. Relief of pain at rest is a useful clue in differentiation.

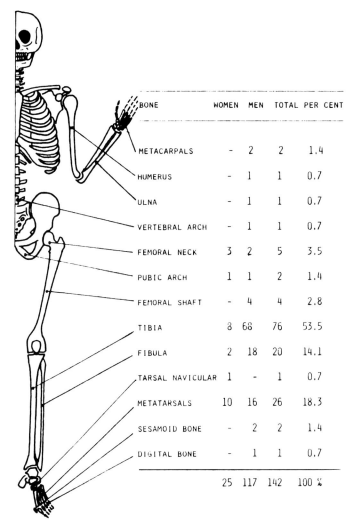

BONE	WOMEN	MEN	TOTAL	PER CENT
METACARPALS	-	2	2	1.4
HUMERUS	-	1	1	0.7
ULNA	-	1	1	0.7
VERTEBRAL ARCH	-	1	1	0.7
FEMORAL NECK	3	2	5	3.5
PUBIC ARCH	1	1	2	1.4
FEMORAL SHAFT	-	4	4	2.8
TIBIA	8	68	76	53.5
FIBULA	2	18	20	14.1
TARSAL NAVICULAR	1	-	1	0.7
METATARSALS	10	16	26	18.3
SESAMOID BONE	-	2	2	1.4
DIGITAL BONE	-	1	1	0.7
	25	117	142	100 %

FIG 1.
Topographic location of stress fractures in athletes. (From Orava S, Puranen J, Ala-Ketola L: Stress fractures caused by physical exercise. *Acta Orthop Scand* 1978; 49:19. Reproduced by permission.)

STRESS REACTIONS

Many individuals who exhibit stress injuries, particularly athletes, undergo radiographic alterations that are limited to areas of periosteal new bone formation, quite different from the fluffy periostitis seen in classic stress fracture (Fig 5). These kinds of alterations have been termed *stress reactions.* Laboratory experiments in dogs have produced reactions of this type without causing any break in the bone's continuity. However, accumulations of osseous microdamage at these sites are not excluded.[18] Floyd and his coworkers[19] have proposed a continuum of stress injury ranging from occult stress reaction to frank stress fracture. This gamut is illustrated in Table 6. On nuclear scanning, these stress reactions may be quite large and therefore may be misconstrued as a large osseous lesion, such as a neo-

TABLE 5.
Fracture Sites Associated With Particular Activities*

Running	Fibula-tibia
Sprint (rare)	
Middle distance	
Long distance	
Hiking	Metatarsal, rare
	Pelvis
Jumping	Pelvis, femur
Tennis	Ulna, metacarpal
Baseball	
Pitching	Humerus, scapula
Batting	Rib
Catching	Patella
Basketball	Patella, tibia, os calcis
Javelin	Ulna
Soccer	Tibia
Swimming	Tibia, metatarsal
Skating	Fibula
Curling	Ulna
Aerobics	Fibula, tibia
Ballet dancing	Tibia
Cricket	Humerus
Fencing	Pubis
Handball	Metacarpal
(Water skiing)	(Pars)

*From McBryde AM: Stress fractures in athletes. *J Sports Med* 1975; 3:212. Reproduced by permission.

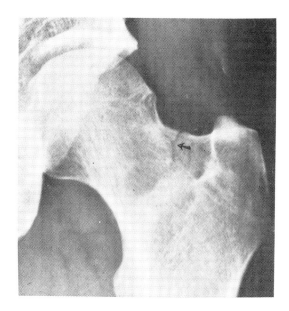

FIG 2.
Type I stress fracture of the femoral neck (*arrow*) in a young soldier showing only a fracture line without reaction.

plasm. Allen[20] has indicated that some stress injuries that appear to be stress reactions may in fact be linear stress fractures with the fracture only being demonstrable by computed tomography (CT). Long-term follow-up of patients with stress reactions will show eventual incorpo-

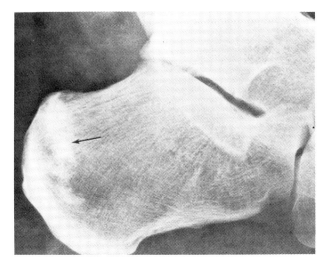

FIG 3.
Type II stress fracture of the calcaneus (*arrow*) in a 22-year-old soldier seen as a band of condensation of bone arranged in a linear configuration.

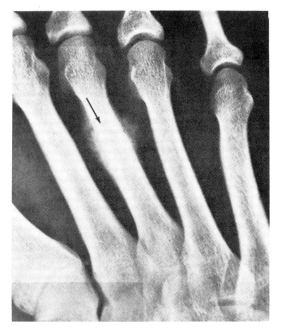

FIG 4.
Type III stress fracture of the third metatarsal (march fracture) seen as an area of exuberant new bone growth with a subtle cleft through the shaft (*arrow*) in a 20-year-old soldier.

ration of the periosteal new bone with no evidence of altered structure of the cortical or medullary bone (Fig 6). There is little doubt that some patients exhibit stress reactions early in their symptomatic period and will eventually go on to frank stress fracture if the stress is not relieved.

NUCLEAR SCANNING

Radionuclide scanning is one of the basic techniques in the diagnosis of stress injury and will yield the earliest confirmatory information. Scanning will detect stress fractures, periosteal reactions, and injuries to soft tissues before radiographic signs become evident.

Clinical experience in the scintigraphic demonstration of stress-related injury has indicated that the three-phase technique is the most useful.[10, 21–23] Using technetium-99m diphosphonate, the area of abnormality will be detected as early as 24 hours after injury because the isotope is incorporated by the osteoblasts in new bone formation.[24] The technique for the three-phase examination is given in Table 7.

Stress injury scintigraphic patterns can be conveniently classified into four grades of bone response according to dimension, extent of the lesion, and tracer concentration in the lesions[25] (Fig 7):

Grade I: Small, ill-defined lesion with mildly increased activity in the cortical region.
Grade II: Larger than grade I, well-defined, elongated lesion with moderately increased activity in the cortical region.
Grade III: Wide fusiform lesion with highly increased activity in the corticomedullary region.

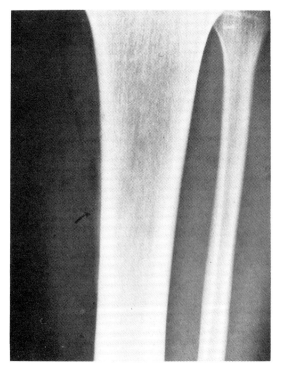

FIG 5.
Stress reaction of the tibia (*arrow*) in a 19-year-old runner.

TABLE 6.

Roentgenologic Findings in the Continuum of Stress Reaction to Stress Fracture*

Continuum of Stress	Pain	Bone Scan	Early Roentgenogram	Delayed Roentgenogram (2–3 wk Later)
Occult stress reaction	Yes	Normal "Hot spot"	Normal	Normal
Stress reaction	Yes	"Hot spot"	Normal or stress reaction	Normal or stress reaction
Occult stress fracture	Yes	"Hot spot"	Normal	Fracture or evidence of healing fracture
Stress fracture	Yes	"Hot spot"	Fracture	. . .

*From Floyd WN Jr, Butler JE, Clanton T, et al: Roentgenologic diagnosis of stress fractures and stress reactions. *South Med J* 1987; 80:433. Reproduced by permission.

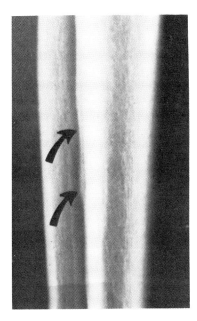

FIG 6.
Old healed stress reaction of the tibia in a 21-year-old athlete (*arrows*).

Grade IV: Wide extensive lesion with intensely increased activity in the transcorticomedullary region.

These gradations reflect the underlying pathologic changes ranging from small stress reactions to frank stress fractures.

Stress injuries are often multifocal and may remain radiographically negative, and it is deemed worthwhile to scan the whole skeleton to detect additional foci.[26] The multiplicity of such foci is often useful in confirming the stress etiology of a given lesion (Fig 8). A related phenomenon has been seen in patients who have undergone surgery for osteosarcoma of the lower limbs and who have demonstrated multiple abnormalities on bone scans in the contralateral limb and upper extremities owing to increased work load on these limbs.[27]

Follow-up studies on patients with stress lesions have shown the minimal time for a fracture to return to normal on bone scan is 5 months. Approximately 90% of the fractures returned to normal by 2 years after injury.[24]

False-positive bone scans are infrequent in the clinical setting of a stress fracture. However, bone cysts, osteomyelitis, sprains, and muscle injuries have produced confusing abnormal scans.[28] Sickle cell disease may also produce an abnormal scan.[10]

TABLE 7.

Technique for Three-Phase Radionuclide Bone Imaging*†

Phase	Tracer Acquisition	Collimation
Dynamic phase: radionuclide angiogram	Bolus IV injection of 20 mCi ^{99m}Tc-MDP, 5.0-sec. images × 8	High or medium sensitivity, parallel hole collimator
Tissue phase: blood pool images	500 k/1,000 k count images	High or medium sensitivity, parallel hole collimator
Metabolic phase: delayed images	500 k/1,000 k count images for high resolution, 500 k/10 min converging 100 k/10 min pinhole	High resolution parallel hole, converging, pinhole collimators

*From Holder LE, Matthews LS: The nuclear physician and sports medicine, in Freeman LM, Weissmann HS (eds): *Nuclear Medicine Annual: 1984.* New York, Raven Press 1984. Reproduced by permission.
†IV = intravenous; ^{99m}Tc-MDP = technetium 99m methylene diphosphonate

STRESS FRACTURE DEVELOPMENT AS SEEN ON BONE SCANS

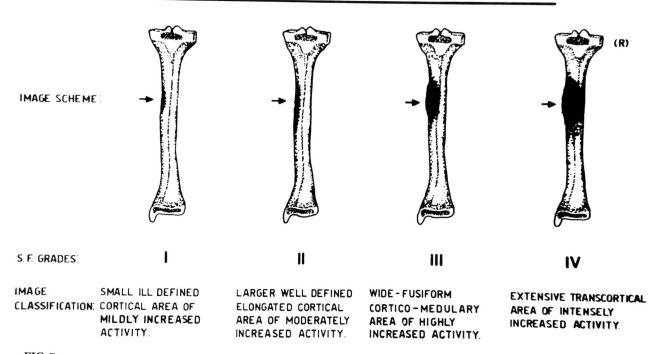

IMAGE SCHEME:

S. F. GRADES: **I**   **II**   **III**   **IV**

IMAGE CLASSIFICATION:

SMALL ILL DEFINED CORTICAL AREA OF MILDLY INCREASED ACTIVITY.

LARGER WELL DEFINED ELONGATED CORTICAL AREA OF MODERATELY INCREASED ACTIVITY.

WIDE-FUSIFORM CORTICO-MEDULARY AREA OF HIGHLY INCREASED ACTIVITY.

EXTENSIVE TRANSCORTICAL AREA OF INTENSELY INCREASED ACTIVITY.

FIG 7.

Four grades (I through IV) of stress fracture evolution as seen on bone scintigraphy presented schematically and in actual bone scintigrams. (From Zwas ST, Elkanovitch R, Frank G: Interpretation and classification of bone scintigraphic findings in stress fractures. *J Nucl Med* 1987; 28:452. Reproduced by permission.)

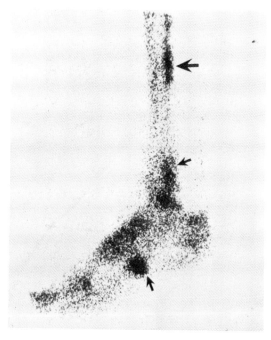

FIG 8.

Nuclear scan of a typical runner's stress fracture of the proximal tibia. Note also multiple other "hot spots" indicating other areas of stress (*arrows*).

COMPUTED TOMOGRAPHY

Conventional linear tomography and polytomography have been usefully employed for many years for better delineation of stress fracture. The advent of CT has added another dimension in that CT has the capacity to view the involved part in cross-section and can confirm the diagnosis of stress fracture when plain radiographic or routine tomographic studies are not diagnostic. Reports of the usefulness of CT in stress injury have indicated that periosteal new bone formation below 1 to 2 mm is not visible. Slight superficial cortical fissures as well as early periosteal callus are better visualized by conventional radiographs than by CT. However, CT reveals endosteal callus even in cases with minor plain film findings,[29] and, when active fracture lines are demonstrated by CT, stress-related injury can be specifically suggested[29-31] (Fig 9) and will help to eliminate difficult differential diagnoses such as neoplasm.

When dealing with insufficiency fractures of the pelvis and particularly the sacrum, CT is essential. These fractures are often impossible to recognize by conventional radiography and may be confused clinically with metastatic disease.[32-35]

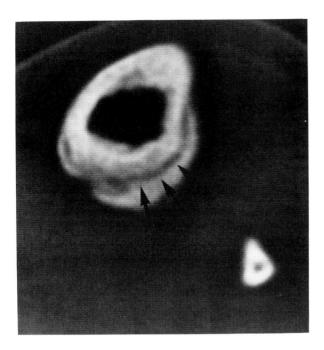

FIG 9.
Computed tomographic scan of the tibia of an 8-year-old girl showing a stress fracture of the lateral cortex (*arrows*).

MAGNETIC RESONANCE IMAGING

Magnetic resonance imaging finds its greatest usefulness in stress injuries in cases where there may be confusion with neoplasm, and, in fact, these are usually the only cases examined. Stress fractures appear as areas of altered signal intensity. Often the changes within the bone marrow are more extensive than anticipated. The signal characteristics will vary with the duration of the injury and are not specific since similar changes may be seen in malignant neoplasia, osteonecrosis, and other disorders.[36,37] The most useful contribution to diagnosis is finding those cases in which the area of altered signal has the configuration of a fracture[38] (Fig 10).

Magnetic resonance examination of asymptomatic professional basketball and collegiate football players showed significant abnormalities in 50% of cases. These abnormalities could have adversely affected scan interpretation in the context of an acute injury. These findings included evidence of ligamentous and tendinous injuries.[39]

THERMOGRAPHY AND ULTRASOUND

Thermography and ultrasound have been advocated as useful techniques for diagnosis of stress fractures.[40, 41]

However, their overall accuracy has not proved to be sufficient to permit their use independently.[42]

Thermography is limited by its inability to differentiate soft tissue from osseous injury. Ultrasound has been used as a technique for stimulating periosteal pain and not as an imaging technique and is therefore subjective. A study comparing radiography, bone scanning, and ultrasound showed bone scanning to yield the best detection rates.[41] There appears to be little justification for the use of either thermography or ultrasound for the diagnosis of stress fracture or reaction in the face of the evidence presented to date.

Ultrasound does have a useful contribution to make to the diagnosis of soft-tissue stress injuries such as tendinous disruption. Such applications will be discussed under the appropriate anatomic sites.

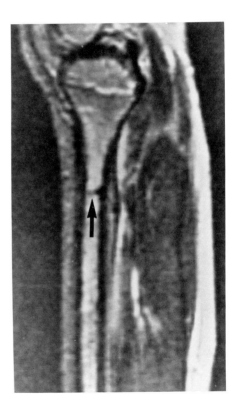

FIG 10.
Magnetic resonance image of the tibia of an 8-year-old girl with a stress fracture (*arrow*) clinically suspected of being a neoplasm. Images made at 1.5 T, T$_2$-weighted sequence, TR$_2$, TE 80 msec.

PART I

Stress Injuries of the Normal Skeleton

2

The Upper Extremity

THE HUMERUS

Stress injuries of the humerus are not common but, when seen, usually occur in adolescents or young adults who are involved in athletics that require a great deal of throwing.

Epiphysiolysis

Epiphysiolysis of the proximal humeral epiphysis was first reported by Dotter in 1953.[43] He first used the term *Little Leaguer's shoulder* to describe the syndrome. Adams[44] reported five additional cases of marked widening of the proximal humeral epiphysis in the throwing arm of adolescent pitchers between the ages of 12 and 15. He felt that this widening represented a traction osteochondrosis. An additional well-illustrated case was reported by Hansen,[45] also in an adolescent baseball player.

These patients complain of persistent anterior shoulder pain when attempting to throw. Radiographic examination in the classic case shows a uniform marked widening of the proximal humeral epiphyseal plate (Fig 11). In some of the previously reported cases, minor fragmentation of the lateral portion of the epiphysis has been noted.[44] Radiographically documented healing of the epiphysis continues beyond the resolution of pain in the shoulder. Bone scanning offers little in diagnosis since the epiphyseal areas are so "hot" normally.

Stress Fracture

Stress fractures of the shaft of the humerus are usually secondary to athletic throwing injuries[8, 46] but have also been implicated in throwing hand grenades[47] and javelins[48] and in arm wrestling[49] and shot-putting[50] (Fig 12). In the cases reported, it appears that the repeated muscular torque on the humerus as a result of throwing led to a stress fracture since many of the subjects reported pain felt in the area for days or weeks before the overt fracture occurred. With continued stress, these injuries converted to overt fractures.

Tug Lesions

I have coined the term *tug lesion* to describe new bone formations at muscle insertions that are the product of the muscular forces on the bone at the point of insertion. These are seen particularly well in the humerus at the insertion of the deltoid muscle and the latissimus dorsi (Figs 13 and 14) and are of no clinical significance.

Osteochondrosis

Osteochondrosis dissecans of the humeral head has been described in a tennis player as a result of overuse.[51] I have seen a Perthes-like aseptic necrosis of the capital humeral epiphysis in an 8-year-old right-handed boy, possibly related to stress (Fig 15).

THE ELBOW JOINT

The elbow joint is the stress focus of many sports and is very prone to injury. These injuries have been studied in detail by Gore et al.[52] They are summarized in Table 8, together with the stresses that precipitate them.

Diffuse generalized stress produces thickening of cortical bone of the humerus and bones of the forearm in adults and overgrowth of ossification centers in the adolescent. The hyperemia caused by overuse prolongs closure of the growth plates. This phenomenon has been particularly well demonstrated in the ossification center of the olecranon in tennis players[53] and baseball players.[54, 55]

Stress injuries to the elbow are particularly common in baseball pitchers and occur during the acceleration phase of pitching as the arm, forearm, and elbow are whipped in

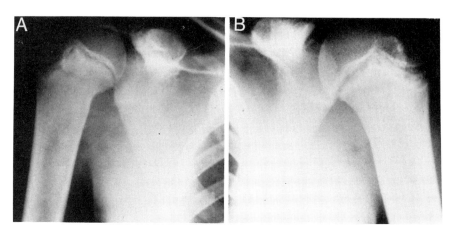

FIG 11.
Comparison views of the normal right shoulder and the symptomatic left shoulder, which demonstrate widening of the epiphyseal plate in a 14-year-old baseball pitcher. (From Hansen NM Jr: Epiphyseal changes in the proximal humerus of an adolescent baseball pitcher: A case report. *Am J Sports Med* 1982; 10:380. Reproduced by permission.)

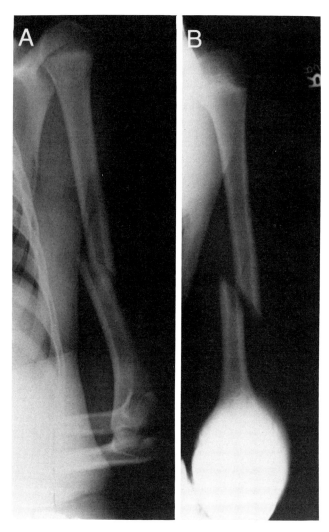

FIG 12.
Stress fracture of the humerus in a 13-year-old Little League pitcher. (From Allen ME: Stress fracture of the humerus: A case study. *Am J Sports Med* 1984; 12:244. Reproduced by permission.)

the direction of the throw. This action forces the radius and ulna into a valgus attitude with respect to the distal humerus, causing distraction at the medial ulnar-humeral articulation and impaction of the radial head into the capitellum. Thus, two major stresses result: medial torsion and lateral compression[52] (Fig 16).

Medial torsion stress is a particular problem for baseball pitchers, tennis players,[56–60] and gymnasts.[61] In the adult, most injuries of this type relate to the pull of the ulnar collateral ligament on the coronal tubercle that produces the ulnar traction spur (Fig 17) and loose bodies. In the adolescent, the injuries are due to the pull of the flexor pronator muscle group at its insertion on the medial epicondylar apophysis. This pull may result in actual avulsion of the apophysis (Fig 18) or chronic traction epiphysitis (Fig 19). In addition to the separation of the ossification center, one may see fragmentation and roughening of the medial epicondyle (Fig 20). As mentioned above, overuse prolongs epiphyseal closure, and the ossification center may remain unfused throughout life[57] (Fig 21). Another reactive change secondary to medial stress is the development of a spur in the lateral margin of the trochlea[58] (Fig 22). This spur has been noted in tennis players as well as javelin throwers. There may also be a loose body in the same location.

Lateral compressive stresses may result in an osteochondral fracture in the adult (Fig 23) as well as in the adolescent (Fig 24). Aseptic necrosis is an uncommon manifestation of chronic lateral stress and usually involves the radial head (Fig 25). In addition, during the adolescent phase of rapid growth and final development of the capitellum, a local avascular necrosis may occur with development of osteochondrosis (Panner's disease) or osteochondrosis dissecans (Fig 26 and 27). The trochlea may be similarly affected.[62]

Extension injuries may be manifested by avulsion of the olecranon process, by acute olecranon apophysitis, or by acute medial epicondylar epiphysitis. The pull of the triceps on its insertion will avulse a portion of the bone and has been reported as a complication of javelin

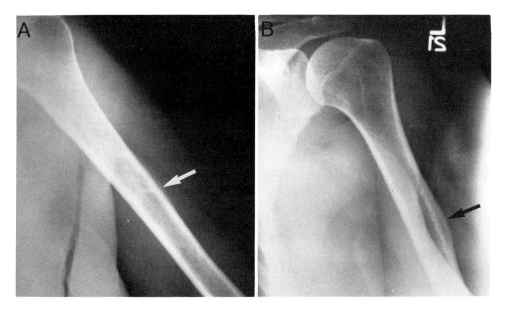

FIG 13.
Two examples of deltoid "tug" lesions, cortical thickenings, and irregularities due to muscle insertion.

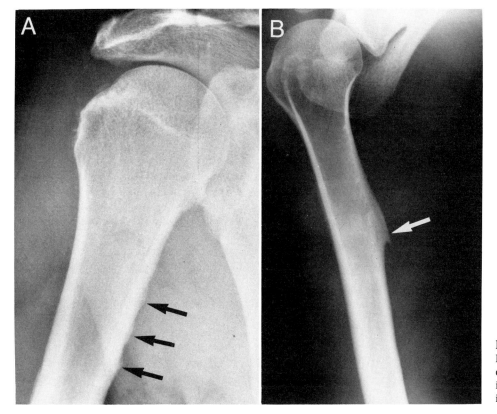

FIG 14.
Latissimus dorsi "tug" lesions, cortical thickenings, and irregularities due to muscle insertion.

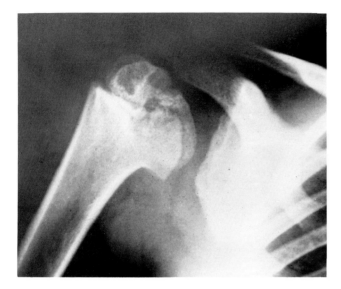

FIG 15.
Aseptic necrosis of the capital humeral epiphysis in an 8-year-old right-handed boy.

throwing[63] (Fig 28). According to Gore et al.,[52] loose body formation is also a manifestation of posterior stress. The articular incongruity created by a hypertrophied olecranon process articulating with a diminutive olecranon fossa may cause synovial shredding and exfoliation of loose bodies at the olecranon tip and in the supratrochlear fossa (Fig 29).

Forceful contraction of the triceps may result in partial avulsion of a portion of the olecranon (Fig 30), requiring surgical reattachment. A similar phenomenon may be seen at the insertion of the biceps muscle with abrupt avulsion or with chronic changes similar to that of myositis ossificans (Fig 31).

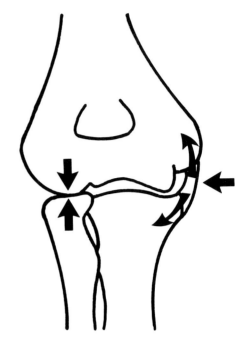

FIG 16.
Valgus stress produces medial distracting stress (*curved arrow*) on ulnar ligament (*single straight arrow*) and lateral radiocapitellar compressive stress (*opposing straight arrows*). (From Gore RM, Rogers LF, Bowerman J, et al. Osseous manifestations of elbow stress associated with sports activities. *AJR* 1980; 134:971. Reproduced by permission).

THE RADIUS

Epiphysiolysis

In adolescence, the distal radial epiphyseal plate represents the weakest part of the anatomy of this bone. Un-

TABLE 8.
Sports-Related Elbow Injuries and Precipitating Stresses*

Type of Stress	Resulting Injuries	
	Adult	Child
Diffuse	Humeral hypertrophy; radial head coronoid process hypertrophy; olecranon process hypertrophy	Hypertrophy and hypermaturity radial head, capitellar, trochlear epiphyses; hypertrophy and hypermaturity medial epicondylar apophysis
Humeral shaft	Spiral fracture humeral shaft	Same as adult
Medial tension	*Acute:* avulsion fracture medial epicondyle; fracture ulnar spur; *chronic:* ulnar traction spur; hyperostosis ulnar groove; loose bodies	*Acute:* avulsion medial epicondylar apophysis; *chronic:* traction apophysitis medial epicondylar apophysis
Lateral compression	*Acute:* osteochondral fracture; *chronic:* loose bodies	*Acute:* osteochondritis dissecans; *chronic:* aseptic necrosis radial head; anterior angulation deformity radial head; loose bodies
Extension	*Acute:* avulsion olecranon process; *chronic:* osteochondral loose bodies	*Acute:* avulsion olecranon apophysitis; *chronic:* traction apophysitis medial epicondylar apophysis

*From Gore RM, Rogers LF, Bowerman J, et al: Osseous manifestations of elbow stress associated with sports activities. *AJR* 1980; 134:971. Reproduced by permission.

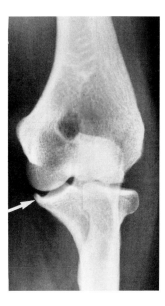

FIG 17.
The ulnar traction spur.

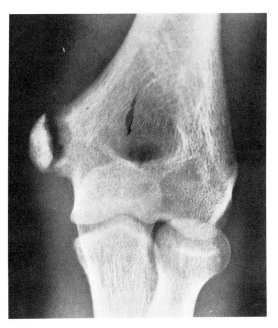

FIG 19.
Stress-induced chronic avulsion of the ossification center of the medial epicondyle in a 17-year-old football passer.

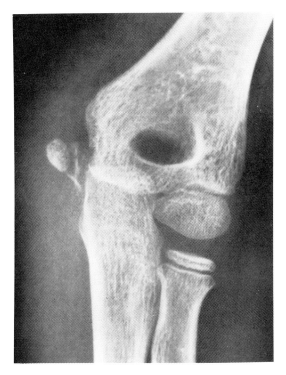

FIG 18.
Stress-induced acute separation of the ossification center of the medial epicondyle in a 10-year-old Little League pitcher (Little Leaguer's elbow).

restricted stress applied to the end of the bone will result in lysis of the plate. In some cases this process will be reflected in radiographic changes including irregular widening of the growth plate, irregular metaphyseal margins, flaring of the metaphyses, and retarded bone maturation. These are actually Salter-Harris type I injuries. These phenomena have been described in adolescent gymnasts[64-67] and roller skaters[68] (Fig 32). Similar changes may involve the distal ulnar epiphysis as well (Fig 33). Long-term follow-up in these patients suggests that residual deformities may persist. A Madelung-like deformity in a gymnast has been reported.[69]

In adolescent weight lifters, a more severe type of stress injury results in fractures of the epiphyseal plates of the radius and ulna with dislocation of the epiphyses.[70] These are essentially Salter-Harris type II fractures (Fig 34). This injury may be accompanied by a torus fracture of the distal ulna as well.

Stress Reactions

Stress reactions due to repetitive physical stress exerted on the bones of the forearm are uncommon but are recognized. A very striking case has been reported by Moss et al.[71] of a mechanic who also weight lifted and slammed his wrists on basketball rims while "dunking the ball." This patient exhibited extensive new bone formation of

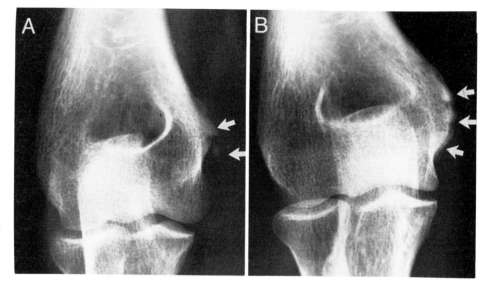

FIG 20.
Traction epicondylitis in a 31-year-old athlete seen as irregularity of the medial epicondyle and ossicle formation. **A,** frontal projection. **B,** oblique projection.

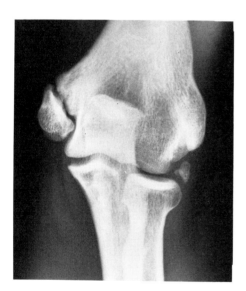

FIG 21.
Ununited epicondylar ossification centers in an 18-year-old athlete.

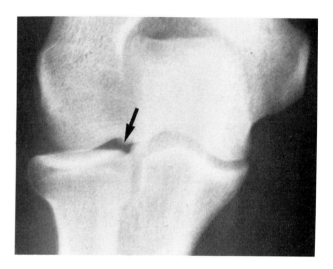

FIG 22.
Lateral trochlear spur (*arrow*) in a tennis player. (Courtesy of Dr. Henry Jones.)

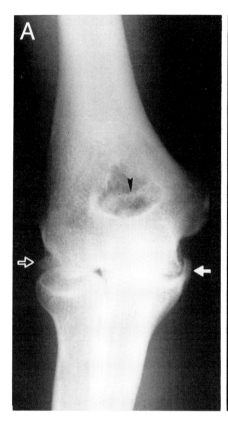
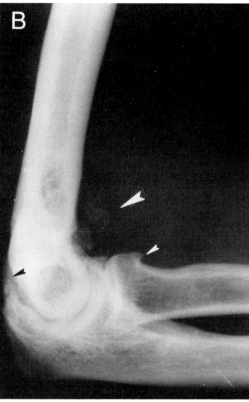

FIG 23.
A 28-year-old professional baseball pitcher with chronic elbow pain, anteroposterior (**A**) and lateral (**B**) views. Cortical hypertrophy of humeral shaft, medial epicondyle, olecranon process and fossa, coronoid process, and radial head squaring (*small white arrowhead*). Ulnar traction spur (*solid arrow*) and anterior (*large white arrowhead*), posterior (*black arrowheads*), and lateral (*open arrow*) loose bodies. (From Gore RM, Rogers LF, Bowerman J, et al: Osseous manifestations of elbow stress associated with sports activities. *AJR* 1980; 134:971. Reproduced by permission.)

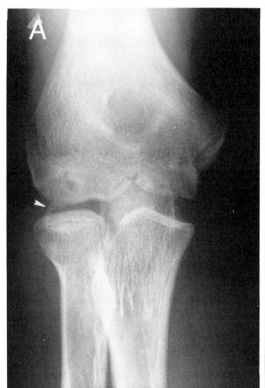
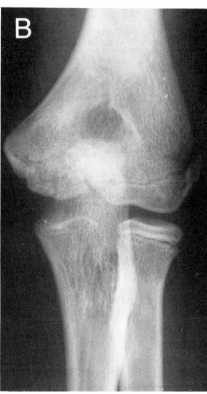

FIG 24.
A 13-year-old avid tennis player with abrupt onset of lateral elbow pain. **A,** anteroposterior view of dominant elbow. Hypertrophy and hypermaturity of elbow epiphyseal and apophyseal growth centers and lateral compartment osteochondral fracture (*arrowhead*) with donor site in capitellum. **B,** nondominant elbow for comparison. (From Gore RM, Rogers LF, Bowerman J, et al: Osseous manifestations of elbow stress associated with sports activities. *AJR* 1980; 134:971. Reproduced by permission.)

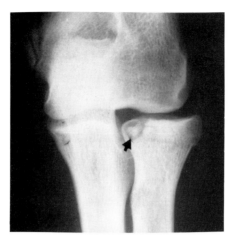

FIG 25.
Osteochondrosis of the radial head (*arrow*) in a 13-year-old athlete.

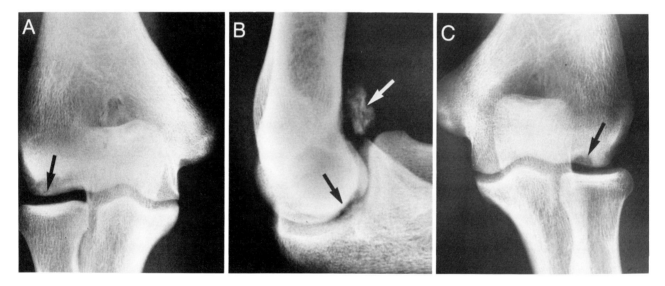

FIG 26.
Bilateral osteochondrosis of the capitellum (*arrows*) in a 15-year-old enthusiastic drummer. **A** and **B,** right elbow. Loose body is seen in lateral projection (*white arrow*). **C,** left elbow.

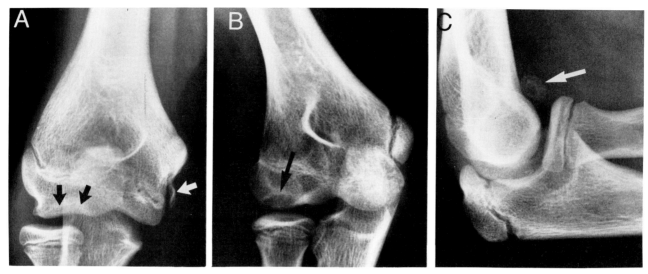

FIG 27.
Stress lesions of the humerus in a 17-year-old baseball player resulting in osteochondrosis dissecans. **A,** frontal projection. Note osteochondritic fossa (*black arrows*) and an extraneous osseous element *(white arrow).* **B,** oblique projection shows the osteochondritic fossa to better advantage (*arrow*). **C,** lateral projection shows osteochondritic loose body within the joint (*arrow*).

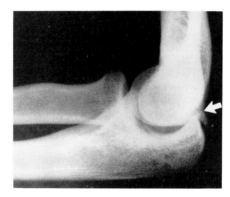

FIG 28.
Irregularity of the tip of the olecranon in an 18-year-old female javelin thrower. (From Miller JE: Javelin thrower's elbow. *J Bone Joint Surg [Br]* 1960; 42:788. Reproduced by permission.)

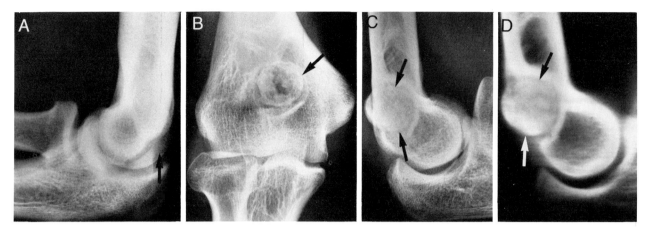

FIG 29.
A, fragment of bone at the olecranon tip in a 21-year-old overhand baseball pitcher. **B,** large loose body in the superior trochlear fossa (*arrow*) in an 18-year-old football player. **C,** lateral projection. **D,** tomogram.

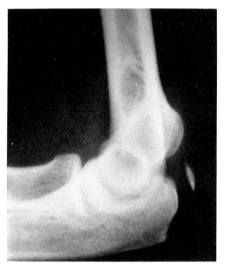

FIG 30.
Avulsion of a portion of the olecranon by forceful pull of the triceps.

FIG 31.
A and **B,** chronic productive changes of biceps avulsion (*arrows*) in a 40-year-old weight lifter.

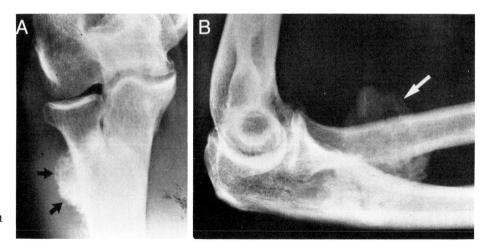

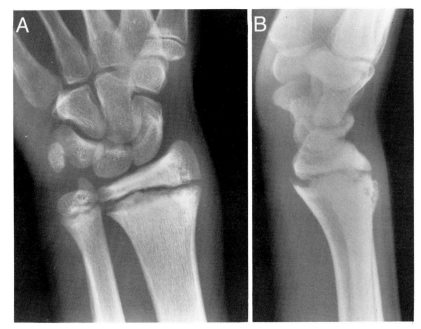

FIG 32.
A, 15-year-old male roller skater. Irregular widening of the radial growth plate. Ill-defined cystic appearances, sclerosis, and flaring of the metaphyses. Flattening of the medial portion of the radial epiphysis. **B,** the widened distal radial growth plate, irregularity, and flaring of the metaphysis and spur formation from the palmar aspect *(arrow).* (From Carter SR, Aldridge MJ, Fitzgerald R, et al: Stress changes of the wrist in adolescent gymnasts. *Br J Radiol* 1988; 61:109. Reproduced by permission.)

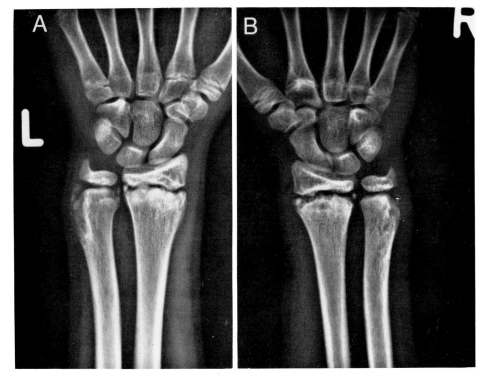

FIG 33.
A 14-year-old male gymnast with irregular widening of both radial and ulnar growth plates and moderate remodeling of the metaphyses. Both ulnae show a lateral zone of metaphyseal expansion, a longitudinal lucency, and an irregular shell of new bone formation. Small fragments are seen at the radial metaphyseal edges bilaterally. (From Fliegel CP: Stress-related widening of the radial growth plate in adolescents. *Ann Radiol* 1986; 29:374. Reproduced by permission.)

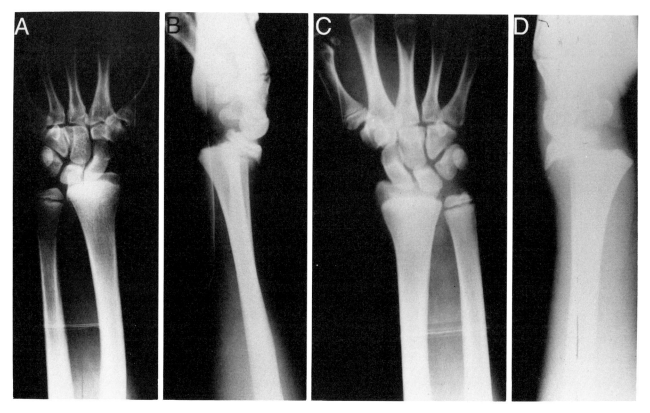

FIG 34.
A 14-year-old boy with bilateral Salter-Harris type II fractures of distal radial epiphysis secondary to weight lifting. (From Gumbs VL, Segal D, Halligan JB, et al: Bilateral distal radius and ulnar fractures in adolescent weight lifters. *Am J Sports Med* 1982; 10:375. Reproduced by permission.)

both forearms without evidence of stress fracture (Fig 35).

A similar, but less marked, example of stress reaction is illustrated in Fig 36. This patient is a 26-year-old employee of an automobile windshield glass replacement company whose daily work required use of a ratchet wrench. The repetitive stress to his forearm resulted in pain and reactive radiographic changes.

Stress Fracture

Frank stress fracture of the radius has been described as being due to a variety of etiologies, all relating to excessive pronation and supination of the forearm. Some of the etiologies include staple fastening,[72] pool playing,[73] carrying heavy objects,[74] and performing bicycle "wheelies,"[75] which involves taking the front wheel of the bicycle off the ground and riding on the back wheel only (Fig 37). A stress fracture of the radius following a fracture of the ulnar diaphysis has been described.[76]

THE ULNA

Stress fractures of the ulna are seen more commonly than those in the radius. Repetitive muscular action of the forearm is the common denominator in all cases. Some of the described etiologies are wheelchair propulsion,[8] use of weight-bearing crutches,[77] pitchfork work,[78] tennis,[79-81] and baseball pitching.[80] Cases of ulnar shaft stress fractures in weight lifters are being reported with increasing frequency[82-84] (Fig 38).

THE WRIST AND HAND

The Hamate

In the past, fracture of the hook of the hamate in athletes has been regarded by some investigators as a stress injury. Studies of large groups of patients have indicated that these injuries are secondary to the direct blow caused by the handle of the tennis racket, golf club, or bat and not by indirect force produced by ligaments or muscles attached to the hook.[85]

The Navicular

A single case of stress fracture of the navicular in a 16-year-old boy has been reported. The patient was a nationally ranked gymnast who practiced 4 to 5 hours per

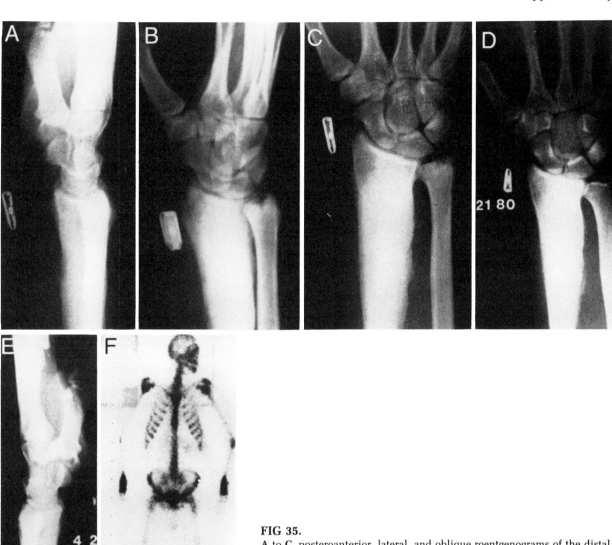

FIG 35.
A to **C,** posteroanterior, lateral, and oblique roentgenograms of the distal ends of
the bones of the right forearm show a dense sclerotic process of the radius with
abundant solid, periosteal reaction incorporated into the cortices. The process is
unusually long, extending from the articular surface of the radius to a point
approximately 5 to 6 cm above the wrist. **D** and **E,** posteroanterior and lateral
roentgenograms of the distal ends of the bones of the left forearm show a
sclerotic process involving the radius, grossly similar in appearance to the
changes in the right radius. **F,** a radioisotope bone scan, using Tc 99m
methylene diphosphonate, obtained at the time of hospital admission in April
1980, demonstrates a striking degree of "uptake" in the distal end of each
radius, corresponding to the sclerotic process demonstrated. The remainder of
the skeleton is normal. (From Moss GD, Goldman A, Sheinkop M: Case report
219. *Skeletal Radiol* 1982; 9:148. Reproduced by permission.)

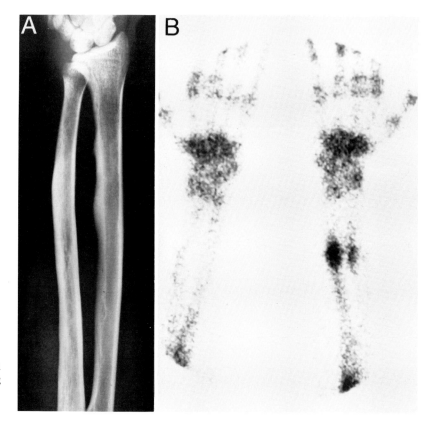

FIG 36.
A, stress reaction of the radius and ulna in a 26-year-old man showing cortical thickening of the radius and ulna. **B,** nuclear scan showing localized area of increased uptake.

day for 6 years. The nuclear scan was abnormal after 8 weeks of symptoms, and the radiologic findings were evident after 5 weeks of plaster immobilization (Fig 39).

The Lunate

Aseptic necrosis of the carpal lunate (Kienböck's disease) is thought to represent a stress-induced injury that leads to devascularization of the major segment of the bone. Support of this concept is lent by the correlation between this disease and ulnar minus variance (ulna shorter than the plane of the adjacent radius). Kienböck's disease is four times more prevalent in patients with ulnar minus variance. Presumably, the ulnar minus variance subjects the lunate to greater compression or shear stress.

The radiologic appearance of the devascularized lunate varies with the duration of the disease, ranging from small fracture lines to sclerosis, collapse, and secondary

fracture of the bone.[86] If untreated, secondary degenerative arthritic alterations occur (Fig 40).

The Pisiform

Stress fracture of the pisiform has been described in volleyball players, possibly secondary to the repetitive impact of the ball as the shooter stops the ball with the flat of the palm.[87] This injury is best seen in the carpal tunnel view.[87]

The Phalanges

Acquired osteolysis of the phalanges has been described in a rather large number of conditions. Among these is mechanical stress. Illustrated in Figure 41 is a single reported case of stress-induced osteolysis of the terminal phalanges of the left hand in an 18-year-old guitar player.[88]

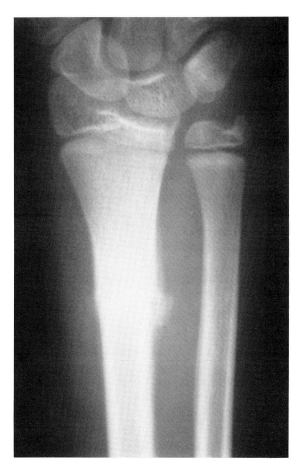

FIG 37.
Stress fracture of the radius in a 12-year-old boy secondary to performing bicycle "wheelies." (From Eisenberg D, Kirchner SG, Green NE: Stress fracture of the distal radius caused by "wheelies." *South Med J* 1986; 79:918. Reproduced by permission.)

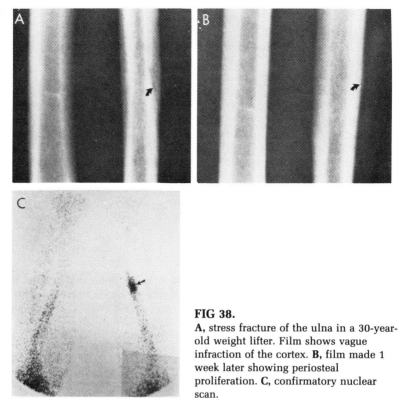

FIG 38.
A, stress fracture of the ulna in a 30-year-old weight lifter. Film shows vague infraction of the cortex. **B,** film made 1 week later showing periosteal proliferation. **C,** confirmatory nuclear scan.

FIG 39.
A, nuclear scan of 16-year-old gymnast shows increased uptake in left navicular bone. **B,** after 5 weeks of plaster immobilization, roentgenogram shows a stress fracture through the waist of the navicular. (From Manzione A, Pizzutillo PD: Stress fracture of the scaphoid waist: A case report. *Am J Sports Med* 1981; 9:268. Reproduced by permission.)

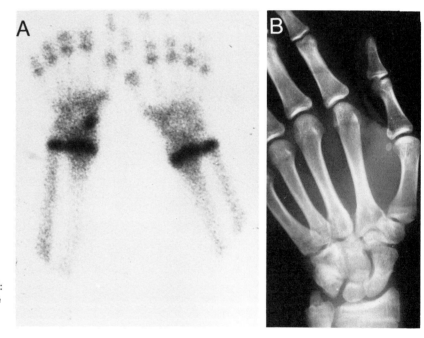

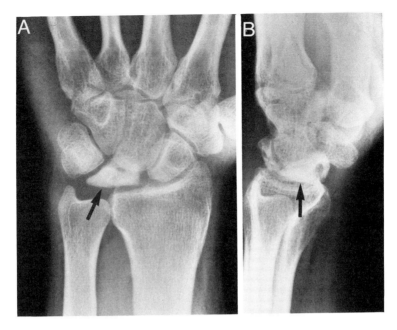

FIG 40.
A and **B**, aseptic necrosis of the lunate (Kienböck's disease).

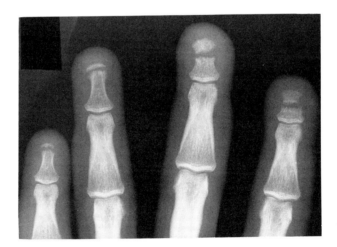

FIG 41.
Stress-induced acro-osteolysis of the terminal phalanges in an 18-year-old guitar player. (From Destouet JM, Murphy WA: Guitar player acro-osteolysis. *Skeletal Radiol* 1981; 6:275–277. Reproduced by permission.)

3

The Shoulder Girdle

THE CLAVICLE

Stress Fracture

Stress fractures of the clavicle are rare. Devas[8] reported a stress fracture in the midshaft as a result of coal shoveling. The ends of the bone as sites of stress injury have been more frequently implicated. Fractures at the medial end are difficult to diagnose in that they are confused with neoplastic processes radiographically. Diagnosis is particularly difficult in the stress fracture of the medial end, which is seen after radical neck dissection [89, 90] and which results from unbalanced muscle traction following surgery (Fig 42). A less complicated case of stress fracture of the medial end has been reported in a 12-year-old boy secondary to carrying heavy books[91] (Fig 43).

Stress Reactions

Osteolysis of the distal end of the clavicle is evidenced by resorption of subchondral bone, and an increased space between the acromion and the clavicle may be seen in patients with repeated stress to the shoulder. The appearance is similar to post-traumatic osteolysis of the clavicle. This entity is seen in persons who perform repeated lifting movements. It has been described in air-hammer operators, weight lifters, and handball players (Fig 44). Cessation of symptoms can produce partial or complete osseous restoration of the clavicle.[92]

Condensing Osteitis of the Clavicle

Condensing osteitis of the clavicle is a term applied to a stress-induced lesion of the medial end of the clavicle characterized by an increase in bone density with sclerosis and, in some cases, hypertrophic changes in the sternoclavicular joint (Fig 45). Most of the reports in the literature suggest that this disease is confined to women,[93, 94] but I have seen a case in a male chiropractor.

As anticipated, nuclear scanning will localize the disease process quite accurately by its increased activity in the medial end of the clavicle, but its pattern has no specificity (Fig 46).

THE SCAPULA

Stress Fracture

Trapshooter's shoulder represents a stress fracture of the coracoid process, here seen in a female trapshooter who shot 200 to 1,000 rounds per week.[95, 96] The fracture was best demonstrated in the axillary projection and healed following cessation of shooting (Fig 47).

Brower et al.[54a] described a stress fracture of the lateral aspect of the scapula at the attachment of the teres minor in an assembly line worker who did repetitive overhead work (Fig 48).

A stress fracture of the infraglenoid tubercle has been reported in baseball players. The injury was located in the posteroinferior body of the glenoid fossa and reflected itself as an exostosis.[97] I suspect that this lesion is more likely a tug lesion at the insertion of the triceps than a stress fracture (Fig 49).

Avulsion Injuries

Avulsion fractures of the scapula result from muscle contraction against a resilient force on the upper limb. Figure 50 illustrates the posterior aspect of the scapula and shows the sites of avulsion fractures reported in the literature. Heyse-Moore and Stoker[56] have reported the most comprehensive series of scapular avulsive lesions and have classified them by location.

Tip of the coracoid
   Avulsion (repeated service at tennis[98]; see Fig 47)
   Resisted muscle pull (fall onto both hands)[99]

30

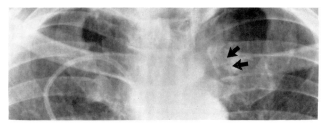

FIG 42.
Stress fracture of the medial end of the left clavicle 2 years
following left radical neck dissection *(arrows).*

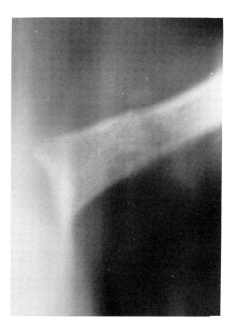

FIG 43.
Tomogram demonstrates a stress fracture of the medial aspect of
the clavicle. (From Kaye JJ, Nance EP Jr, Green NE: Fatigue
fracture of the medial aspect of the clavicle. *Radiology* 1982;
144:89. Reproduced by permission.)

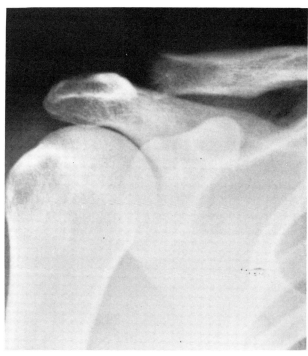

FIG 44.
Stress-induced osteolysis of the distal end of the clavicle in a
23-year-old man who was employed as a package handler who
lifted heavy weights regularly.

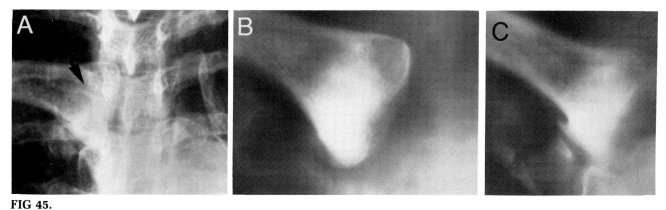

FIG 45.
Condensing osteitis of the right clavicle *(arrow)* in a 32-year-old woman who is employed as a physical therapist. **A,** plain film. **B,**
tomogram showing sclerosis of the medial end of the right clavicle. **C,** tomogram showing spurring of the inferior aspect of the right
clavicle.

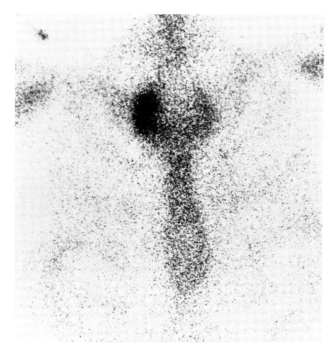

FIG 46.
Nuclear scan of the patient shown in Figure 45 demonstrating increased uptake at the medial end of the right clavicle.

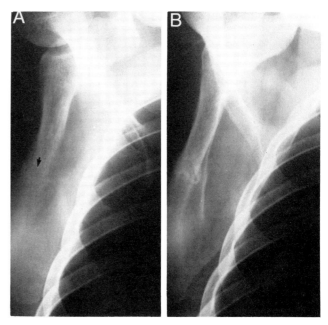

FIG 48.
Stress fracture of the lateral margin of the scapula. **A,** poorly defined linear lucency (*arrow*). **B,** film 2 weeks later shows more abundant periosteal reaction. Linear lucency has filled in partially and is better defined. (From Brower AC, Neff JR, Tillema DA: An unusual scapular stress fracture. *AJR* 1977; 129:519. Reproduced by permission.)

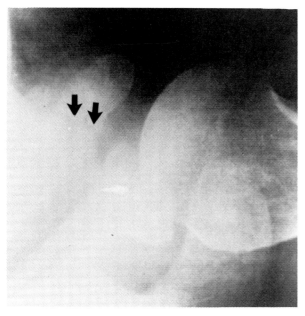

FIG 47.
Stress fracture of the coracoid process (*arrows*) secondary to trapshooting. (From Sandrock AR: Another sports fatigue fracture. *Radiology* 1975; 117:274. Reproduced by permission.)

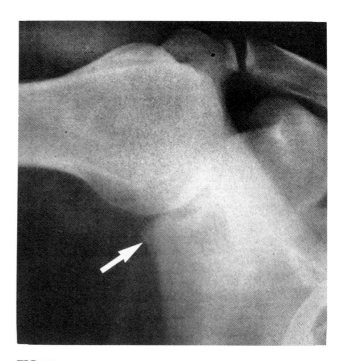

FIG 49.
Probable stress fracture of the infraglenoid tubercle in a baseball pitcher (*arrow*), which resolves as an exotosis. (From Bennett GE: Shoulder and elbow lesions of the professional baseball pitcher. *JAMA* 1941; 117:510. Reproduced by permission.)

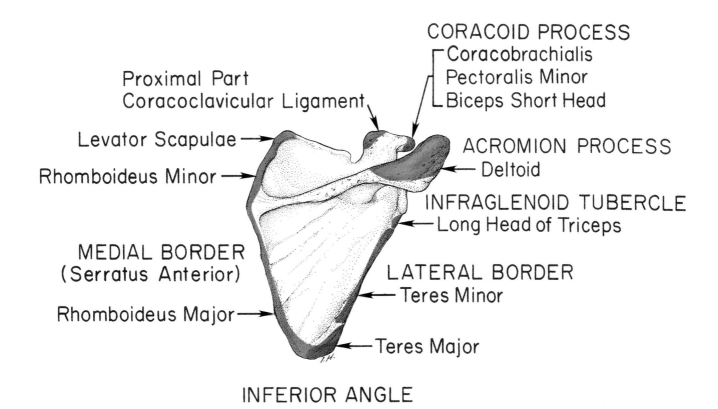

CORACOID PROCESS
⎡Coracobrachialis
⎢Pectoralis Minor
⎣Biceps Short Head

Proximal Part
Coracoclavicular Ligament

Levator Scapulae

Rhomboideus Minor

ACROMION PROCESS
— Deltoid

INFRAGLENOID TUBERCLE
— Long Head of Triceps

MEDIAL BORDER
(Serratus Anterior)

Rhomboideus Major

LATERAL BORDER
— Teres Minor

— Teres Major

INFERIOR ANGLE

FIG 50.
Sites of avulsion fractures of the scapula and the responsible
muscles.

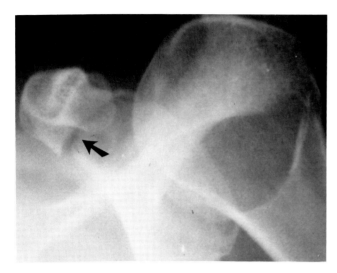

FIG 51.
Avulsion fracture of the base of the coracoid (*arrow*) following a toboggan accident. (From Heyse-Moore GH, Stoker DJ: Avulsion fractures of the scapula. *Skeletal Radiol* 1982; 9:27. Reproduced by permission.)

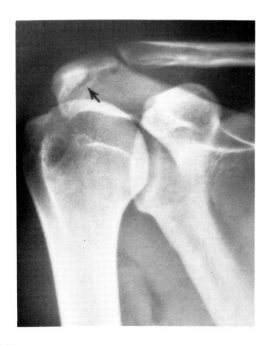

FIG 52.
Fracture of the acromion (*arrow*) following a motor car accident with no direct trauma to the shoulder.

Body of the coracoid (Fig 51)
  Resisted muscle pull (fall down stairs)[100]
  Electroconvulsive therapy[101, 102]
  Avulsion by coracoid ligament[9, 56]
Acromion process (Fig 52)
  Resisted muscle force (using screwdriver with arm above head)[103]
  Car accident[56]

Infraglenoid tubercle (Fig 53)
  Electroconvulsive therapy[101]
  Resisted muscle pull (car accident and football injury)[56]
Lateral border (Fig 54)
  Resisted muscle pull (fall on outstretched arm and toboggan accident)[56]
Inferior angle (Fig 55)
  Electroconvulsive therapy[101]
  Resisted muscle pull (fall on outstretched arm and toboggan accident)[56]
Superior border (Fig 56)
  Resisted muscle pull by the omohyoid and levator muscles at their insertions[104, 105]

The common denominators in all these injuries are uncoordinated muscle contracture, resisted muscle pull, or avulsion of a ligamentous attachment.

VASCULAR LESIONS

There are two vascular lesions of the shoulder girdle that have been implicated in musculoskeletal stress. The first of these is the so-called effort thrombosis of the subclavian vein that is supposedly due to compression of the vein between the clavicle anteriorly and the first rib posteriorly. The syndrome is characterized by swelling of the dominant arm with distention of the superficial veins and some cyanosis. Venography is helpful in demonstrating the point of obstruction[106] (Fig 57).

The second lesion is the quadrilateral space syndrome, which is caused by occlusion of the posterior humeral circumflex artery in the quadrilateral space. The quadri-

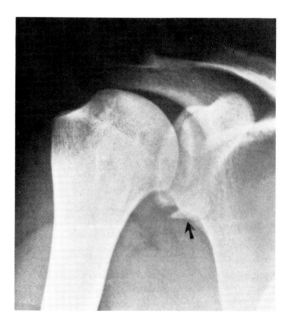

FIG 53.
Fracture of the infraglenoid tubercle in a 20-year-old wrestler (*arrow*).

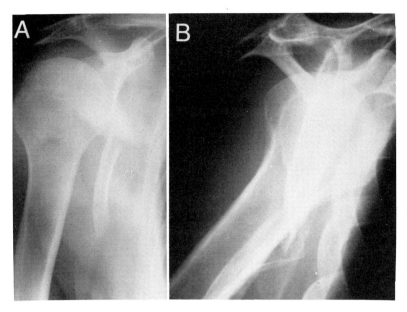

FIG 54.
A and **B**, radiographs of the shoulder showing avulsion fracture of the lateral border and inferior angle of the scapula. (From Heyse-Moore GH, Stoker DJ: Avulsion fractures of the scapula. *Skeletal Radiol* 1982; 9:27. Reproduced by permission.)

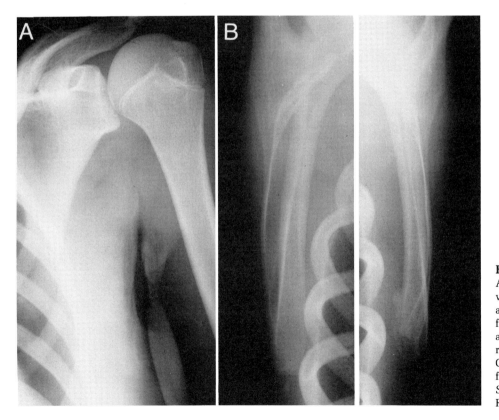

FIG 55.
A and **B**, frontal and axial views of the left scapula (with axial view of the right scapula for comparison) showing avulsion and periosteal reactions. (From Heyse-Moore GH, Stoker DJ: Avulsion fractures of the scapula. *Skeletal Radiol* 1982; 9:27. Reproduced by permission.)

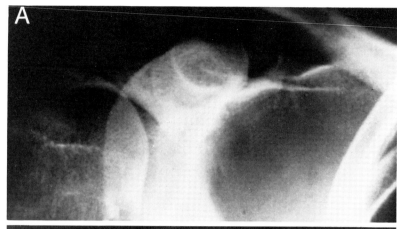

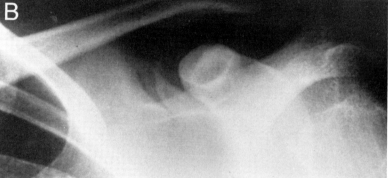

FIG 56.
A and **B,** avulsion fractures of the superior margins of the scapulae following motor vehicle accident. (From Williamson DM, Wilson-MacDonald JW: Bilateral avulsion fractures of the cranial margin of the scapula. *J Trauma* 1988; 28:713. Reproduced by permission.)

FIG 57.
Venogram in abducted position showing obstruction of subclavian vein at medial border of the scapula. (From Wright RS, Lipscomb AB: Acute occlusion of the subclavian vein in an athlete: Diagnosis, etiology and surgical management. *J Sports Med* 1975; 2:343. Reproduced by permission.)

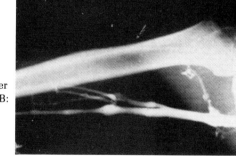

lateral space is located over the posterior scapula and the subdeltoid region. It consists of the teres minor superiorly, the long head of the triceps medially, the teres major inferiorly, and the surgical neck of the humerus laterally. Symptoms consist of paresthesia in the upper extremity and pain poorly localized to the anterior aspect of the shoulder in an indeterminate distribution. The diagnosis is usually made at arteriography, which shows occlusion of the posterior humeral circumflex artery when the arm is abducted and externally rotated (Fig 58). This syndrome is seen in athletes who throw, for example baseball pitchers.[107]

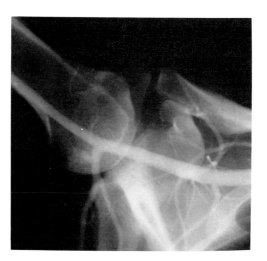

FIG 58.
Subclavian arteriogram with the arm in extreme abduction and extreme rotation. There is complete occlusion of the posterior humeral circumflex artery. (From Cormier PJ, Matalon TAS, Wolin PM: Quadrilateral space syndrome: A rare cause of shoulder pain. *Radiology* 1988; 167:797. Reproduced by permission.)

4

The Thoracic Cage

THE RIBS

Stress Fractures

Stress fractures of the ribs were extremely common after thoracoplasty, but as this technique is no longer used, they are no longer seen in this association.[8] Probably the most common cause of stress fractures of the ribs is violent or sustained coughing. The cough fracture takes place because of a shearing stress brought about by the action of two sets of opposing muscles that interdigitate on the ribs.[108, 109] The break occurs where the serratus muscle and costal slips of the latissimus dorsi interdigitate with fibers of the external oblique muscle. Anatomically, this point corresponds to the middle third of the ribs, and this is the most common site of the fracture. Cough fractures have been reported in every rib, including the first rib.[110]

Cough fractures of the lower ribs have been noted in pregnant women late in their pregnancies.[111–113] The sixth to tenth ribs are most commonly injured, usually in the posterior axillary line.

In my experience, cough fractures are most commonly seen in asthmatics, and I have seen several patients in whom images of the repairing rib fractures have been mistaken for multiple parenchymal lesions, including metastatic deposits (Fig 59).

Stress fractures of the first rib are the result of a number of types of repetitive activities, including coughing,[110] repetitive lifting, raking,[114, 115] baseball pitching,[116, 117] rebounding in basketball,[118] and backpacking (Fig 60), and as a complication of midline sternotomy.[119]

Other athletic pursuits have been implicated in stress fracture of the ribs. Golfers, particularly beginners who are trying hard to master the proper technique, are prone to develop stress fractures in the left posterior ribs when the player is right-handed and in the left upper ribs when the player is left-handed.[120]

A similar phenomenon has been described in the lower ribs of female rowers along the posterolateral segment of the rib where the bending stresses are the greatest.[121] Tennis has also been implicated as an etiology.[8]

Stress Reactions

Macones et al.[122] have described a group of patients with hyperostotic changes in the ribs and spine, mostly on the right side. These changes have related to heavy work, employment, and handedness (Fig 61).

THE STERNUM

Stress fractures of the sternum have been described in elderly osteoporotic patients (see Chapter 8) but are most unusual in individuals with normal skeletons. One of the reported cases occurred during exercise in a 16-year-old boy with cystic fibrosis and another in a 19-year-old wrestler whose symptoms started about a year before the onset of severe sternal pain. Radiography demonstrated the fracture, which was confirmed by bone scanning (Fig 62).

Stress fractures of the sternum are not clinically significant in themselves, but they may simulate more serious clinical conditions such as myocardial infarction or pulmonary embolism.

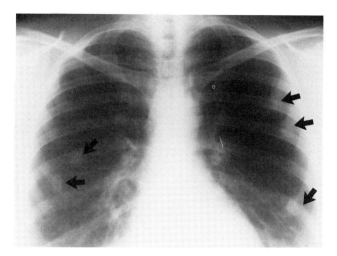

FIG 59.
Multiple cough fractures (*arrows*) in a 20-year-old asthmatic.

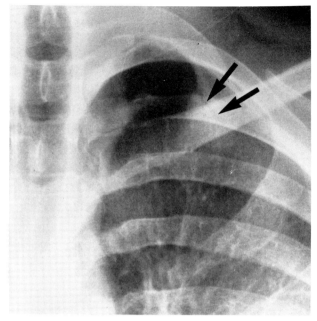

FIG 60.
Stress fracture of the first rib (*arrow*) in a 20-year-old backpacker.

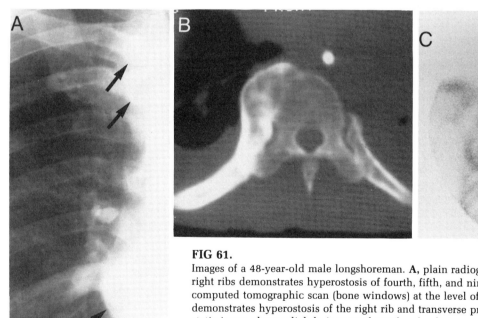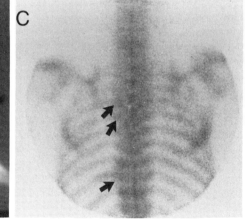

FIG 61.
Images of a 48-year-old male longshoreman. **A,** plain radiograph of medial aspect of right ribs demonstrates hyperostosis of fourth, fifth, and ninth ribs (*arrows*). **B,** computed tomographic scan (bone windows) at the level of the ninth vertebral body demonstrates hyperostosis of the right rib and transverse process. **C,** Tc 99m posterior static image shows slightly increased uptake of radiotracer (*arrows*) at the fourth, fifth, and ninth levels on the right. (From Macones AJ Jr, Fisher MS, Locke JL: Stress-related rib and vertebral changes. *Radiology* 1989; 170:117. Reproduced by permission.)

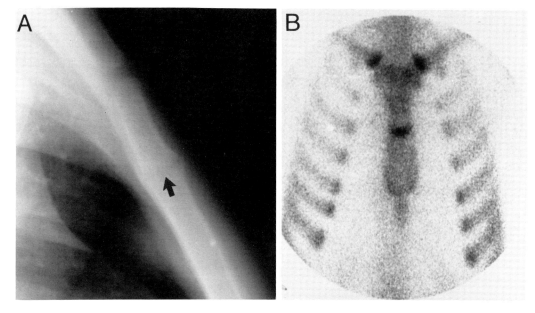

FIG 62.
A, lateral view of the sternum showing a stress fracture (*arrow*) in a 19-year-old wrestler. **B,** bone scan confirms the presence of the fracture. (From Keating TM: Stress fracture of the sternum in a wrestler. *Am J Sports Med* 1987; 15:92. Reproduced by permission.)

5

The Spine

Stress Fractures

Fracture of the spinous processes of the lower cervical and upper lumbar vertebrae is commonly called the clay-shoveler's fracture since it may be seen in men who are shoveling heavy soil or clay (Fig 63). These fractures of the cervical spine are believed to be the product of repeated stress caused by the pull of the trapezius and rhomboid muscles on the spinous process.[123, 124]

Stress fractures of the rigid spine of ankylosing spondylitis occur most commonly at the lower thoracic and upper lumbar spines (Fig 64). The mechanical response of the rigid spine is comparable to that of a long bone, and stress fractures here behave comparably.[125] Similar lesions may be seen as the result of direct trauma but usually do not demonstrate as much sclerotic reaction,[126] at least not until pseudarthroses develop.[127]

The relationship between musculoskeletal stress and spondylolysis has been well documented in recent years. Although it is recognized that spondylolysis may be present at birth[128] and may result from acute trauma,[129] there is good evidence that most cases of spondylolysis are stress induced. These lesions may be unilateral or bilateral and are most commonly seen in gymnasts.[61, 130–132] The fractures usually occur in the pars interarticularis and result from repetitive stress. Although it can occur in any sport, this type of stress fracture predominates in sports involving highly competitive and strenuous training programs such as gymnastics and football.[61, 133, 134] Such injuries have been described in young children as well.[135, 136]

The athlete with stress-induced spondylolysis complains of low back pain localized to one or both sides of the midline, aggravated by twisting and hyperextension of the lumbar spine. Early recognition of this lesion is important. The fracture will heal with rest but, if undetected, may result in permanent nonunion[133, 136] (Fig 65).

Initial radiographs, including oblique views, may be normal in athletes in whom stress fracture of the pars in-terarticularis is developing. Tc 99m-polyphosphate bone scanning is a valuable aid in early diagnosis, often demonstrating a stress lesion before it can be detected radiologically or when the initial bone resorptions are starting to occur (Fig 66).

The bone scan can also be useful in indicating whether the stress fracture is recent or old, which has implications for treatment. It is thought that a normal scan excludes significant active bone pathology and raises the possibility of an alternative diagnosis.[137, 138]

Stress fractures of the sacrum are frequent in osteoporotic women but are most unusual in young, vigorous subjects. An example of such a case is shown in Figure 67, a sacral stress fracture in a runner.

Stress Reactions

Sclerotic pedicles were once considered highly indicative of the presence of an osteoid osteoma. More recent experience indicates that stress reaction is the most common etiology. In patients with unilateral spondylolysis, there is often a stress-induced hypertrophy of the lamina and pedicle of the contralateral side of the same vertebra.[139] This unilateral hypertrophy appears to represent a physiologic reaction to stress on an unstable neural arch. It is important that this underlying mechanism of production of the sclerotic pedicle be understood since surgical excision of the sclerotic pedicle can be expected to cause painful instability[140] (Figs 68 and 69).

Occasionally, sclerotic pedicles may be seen in response to a neural arch defect in the vertebra below (Fig 70). Pedicle hypertrophy may also result from hypoplasia of the pedicle of the opposite side (Fig 71).

Patients with scoliosis may also demonstrate hypertrophy and sclerosis of pedicles, usually on the concave side of the scoliosis as a result of the stress of faulty weight bearing (Fig 72). Some patients with scoliosis may show

41

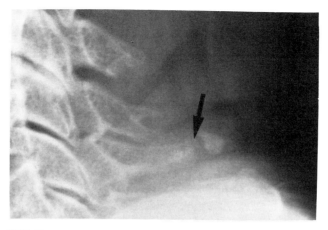

FIG 63.
Clay shoveler's fracture of the spinous process of C-6 (*arrow*) in
a 55-year-old laborer.

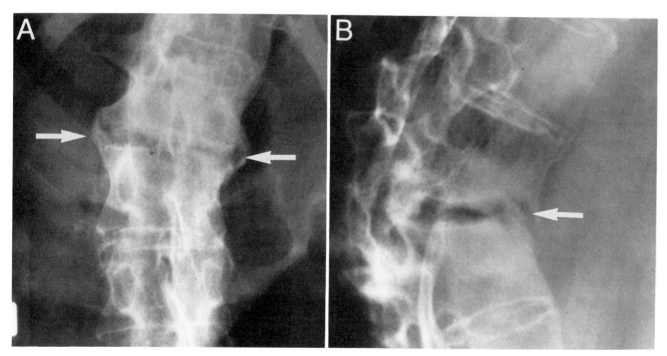

FIG 64.
A and **B**, stress fracture of L-1 and L-2 (*arrows*) in a 55-year-old man with ankylosing spondylitis.

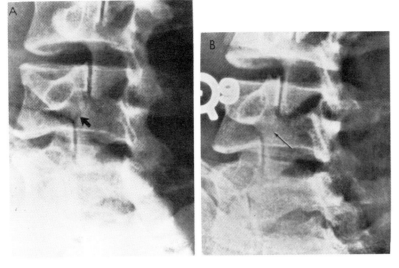

FIG 65.
A, stress-induced spondylolysis of the pars interarticularis of L-4 in a 20-year-old gymnast. **B,** follow-up film obtained 6 months later shows complete healing.

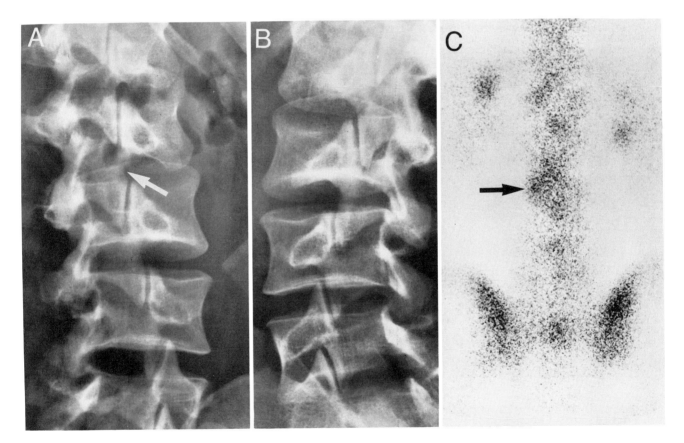

FIG 66.
A, incipient osteolysis of the pars interarticularis of L-2 on the left side in a 21-year-old football player *(arrow)*. **B,** comparison view on opposite side. **C,** bone scan shows equivalent area of increased uptake *(arrow)* in PA view.

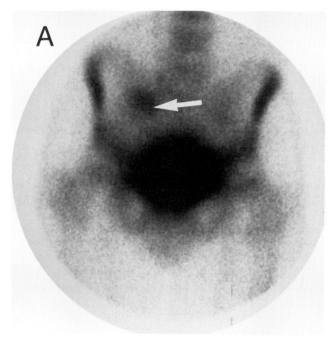
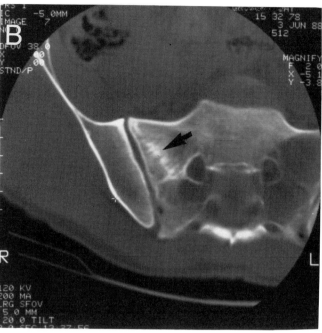

FIG 67.
Sacral stress fracture in a 26-year-old competitive long-distance runner. **A,** scintigram (frontal projection) of pelvis shows increased uptake of radionuclide in right sacral area (*arrow*). **B,** axial computed tomogram of sacrum shows sclerotic area adjacent to right sacroiliac joint (*arrow*). (From Czarnecki DJ, Till EW, Minikel JL: Unique sacral stress fracture in a runner. *AJR* 1988; 151:1255. Reproduced by permission.)

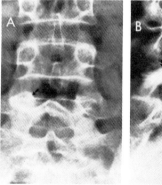
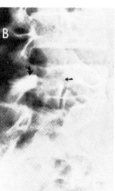
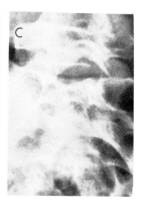

FIG 68.
Sclerotic pedicle sign in a 14-year-old female gymnast. **A,** sclerotic pedicle in frontal projection. **B,** left posterior oblique projection. Small arrow indicates spondylolysis of the pars interarticularis on the patient's left side. Large arrow on the left indicates the sclerotic pedicle. **C,** right posterior oblique projection showing an intact neural arch on the patient's right side.

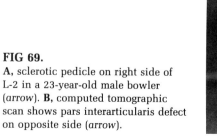
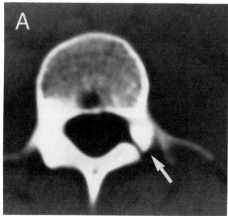
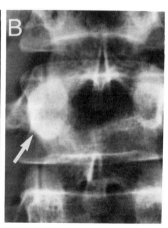

FIG 69.
A, sclerotic pedicle on right side of L-2 in a 23-year-old male bowler (*arrow*). **B,** computed tomographic scan shows pars interarticularis defect on opposite side (*arrow*).

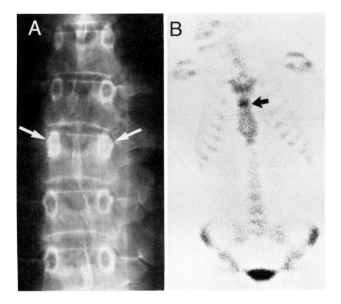

FIG 70.
A, bilateral sclerotic pedicles in a mid-dorsal vertebra (*arrows*) in a patient with an incomplete neural arch in the vertebra below. **B,** corresponding nuclear scan.

Osteochondrosis

Juvenile kyphosis (adolescent spondylodystrophy, Scheuermann's disease) has long been regarded as a form of osteochondrosis of the ring apophysis of the spine during adolescence when these apophyses are ununited to the vertebral body. The disease leads to irregularities of the margins of the bodies, wedging of the bodies, narrowing of the intervertebral discs, and Schmorl's node deformities. The disease predominates in males (Fig 75).

Recent work has indicated that stress may be an important etiology of this disease. A report by Horne et al.[141] on spinal column damage from water ski jumping and a study comparing Danish army recruits from farming areas with those from Copenhagen[142] have shown a decided increase in incidence of this disease in those youngsters whose spines were subjected to axial compressive forces. Similar changes have been seen with other activities that produce similar forces. These activities include horseback riding,[143] snowmobiling over rough ground,[144] and gymnastics.[145] Figure 76 shows Scheuermann-like changes in the lumbar spine of a 38-year-old equestrienne who has been involved in horse jumping since the age of 12.

Hypertrophic changes in the ribs and vertebrae secondary to heavy work-related employment and handedness are discussed on page 39 and illustrated in Figure 61.

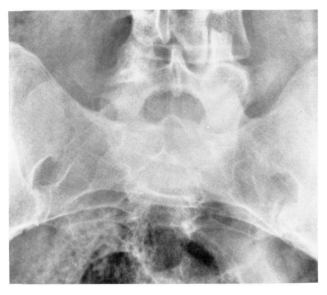

FIG 71.
Hypertrophy of the pedicle on the left side of L-5 secondary to hypoplasia of the right pedicle.

stress sclerosis of pedicles and increased nuclear uptake on both the concave and convex sides of the scoliosis, but at different levels (Fig 73). Similarly, asymmetry of the sacroiliac joint apparently can also produce abnormal stresses in the lumbar spine, thus causing a sclerotic pedicle (Fig 74).

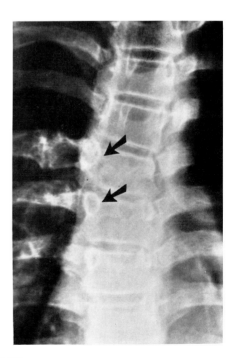

FIG 72.
Sclerotic and enlarged pedicles resulting from scoliosis.

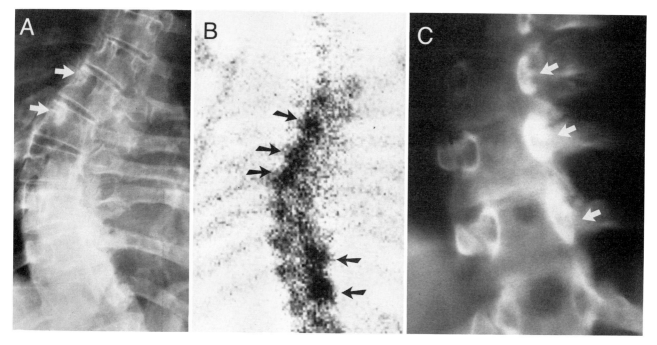

FIG 73.
Stress-induced sclerotic pedicles in a 33-year-old scoliotic woman. **A,** frontal film shows sclerotic pedicles on the convex side of the curve (*arrows*). **B,** nuclear scan shows increased uptake on both the concave and convex sides of the curve, but at different levels (*arrows*). **C,** tomogram shows sclerotic pedicles on the concave side of the curve (*arrows*).

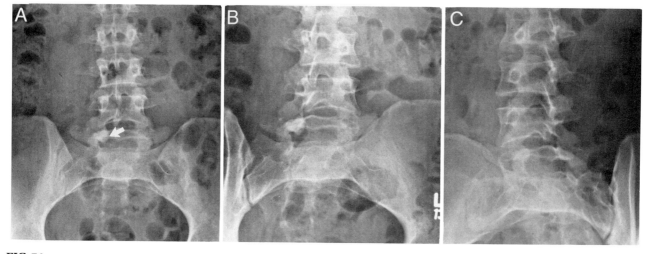

FIG 74.
A, sclerotic pedicle on the right side of L-5 (*arrow*) in a 20-year-old woman. Note apparent absence of left sacroiliac joint. **B** and **C,** oblique views show that both sacroiliac joints are present but apparently in different planes.

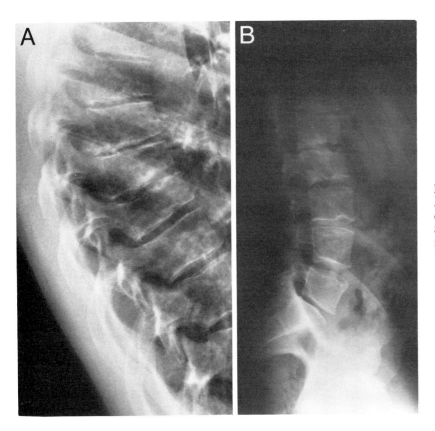

FIG 75.
A, typical Scheuermann's disease of the
dorsal spine in a 25-year-old man. **B,** healed
Scheuermann's disease of the lumbar spine
in a 22-year-old man.

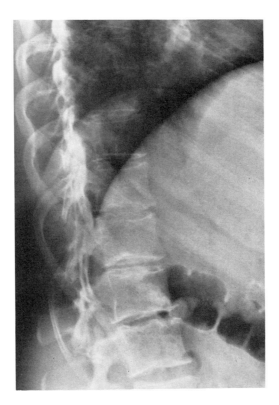

FIG 76.
Lower dorsal and lumbar vertebrae end-plate irregularity,
vertebral wedging, and disc narrowing in a 38-year-old
equestrienne.

6

The Pelvis

Stress Fractures

Stress fractures of the pelvis are most often the insufficiency type seen in postmenopausal women, and this subject is discussed in Chapter 8. However, stress fractures of the inferior ramus of the pubis occur in normal children and adults. In adults, such fractures may also occur in the superior ramus of the pubis or in the body of the pubic bone near the hip.

It is important to note that most stress fractures of the pelvis occur at the site of the ischiopubic synchondrosis. In children, the normally "swollen" synchondrosis should not be mistaken for a stress fracture, even if it is seen unilaterally[146] (Fig 77).

According to Morris and Blickenstaff,[7] pubic stress fractures are probably related to muscle activity rather than weight bearing alone. The fracture occurs between the insertions of two opposing muscle groups. The medial group includes the adductor longus and brevis, the gracilis, and the medial portion of the adductor magnus. These muscles adduct, flex the thigh, and rotate it medially. The lateral group consists of the lateral portion of the adductor magnus and the hamstrings, which extend the thigh and rotate it laterally. The stress fracture may be the result of the antagonistic action of these two massive groups of muscles.

Groin pain is the usual presenting complaint, and there is tenderness to palpation over the involved bone. Healing invariably occurs with limitation of activity.

Stress fracture of the pubis is quite common in military recruits[7, 147, 148] and runners, particularly those who engage in marathon running[149, 150] (Fig 78). It has also been described in swimmers[151] and in pregnant women.[152] The mechanism for these fractures in pregnancy is not clear.

Stress Reactions

Since it is seen most commonly in pregnant women, osteitis condensans ilii is generally considered to be the result of muscular imbalance with resulting stress on the sacroiliac joint[153] (Fig 79). It is a uniform unilateral or bilateral density in the medial portion of the iliac bone adjacent to the sacroiliac joint. The entity is largely a radiologic one since its relationship to back symptoms has not been confirmed. The condition resolves spontaneously in most cases and is not commonly seen in the elderly.

A similar type of sclerotic reactive change combined with bone resorption may be seen in the symphysis pubis in patients with instability of the pelvic ring. Such instability is seen more commonly in women who have borne children but may also be seen in those who have not (Figs 80 and 81). In these patients the symphyseal changes result from laxity of the symphysis pubis and sacroiliac joints, which leads to abnormal mobility and stress alteration in the symphysis. This phenomenon can be demonstrated by filming the pelvis with the patient standing on one leg and then the other. Comparison of these two films will show abnormal mobility of the symphysis pubis[154, 155] (Fig 82).

Tug Lesions

In middle-aged and elderly patients, there is often marked spurring of bone at points of muscle insertion as a reaction to muscle pull at these sites. Figure 83 shows this type of reaction in an elderly woman.

Avulsion Fractures

The growth of the bony pelvis is completed by multiple apophyseal centers that serve as the origin or insertion of several muscle groups. Sudden or chronic forces applied by these muscles will result in avulsion injury. The cartilaginous junction of the apophysis and the parent bone represents the weakest part of the skeleton in the adolescent. The symptoms of avulsion may be sudden in onset, but careful review of the patient's history and radiographs

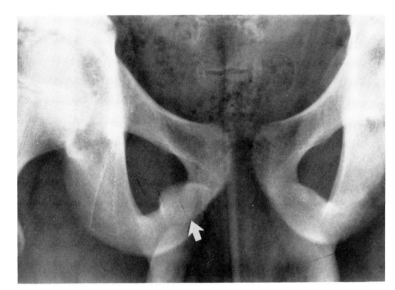

FIG 77.
Normal unilateral "swollen" ischiopubic synchondroses in a 12-year-old boy (*arrow*), not to be mistaken for a stress fracture.

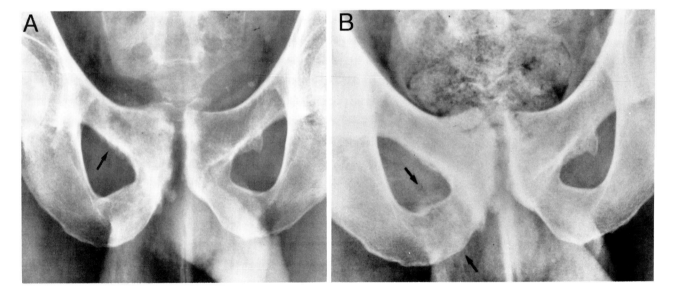

FIG 78.
A, stress fracture of the right pubis in a 38-year-old marathon runner (*arrow*). Note also reactive changes in the symphysis pubis. **B,** the patient continued running and subsequently developed stress fracture of the inferior pubic ramus (*arrows*).

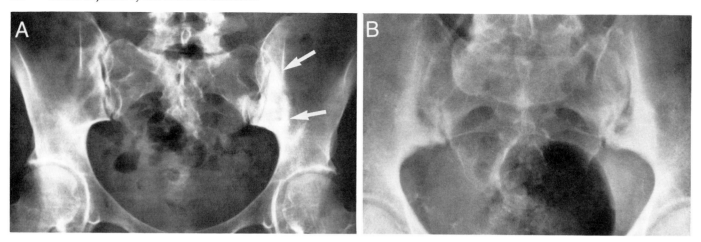

FIG 79.
A, left osteitis condensans ilii (*arrows*) in a 35-year-old woman. **B,** bilateral condensans ilii in a 60-year-old woman.

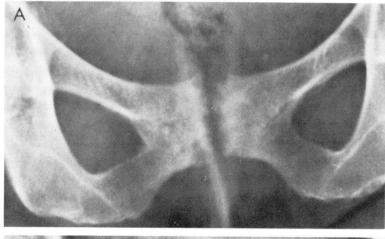

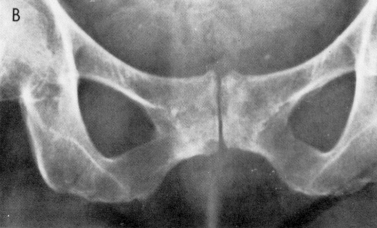

FIG 80.
A, sclerosis and fragmentation of the symphyseal margins in a 40-year-old woman, an active tennis player, who complained of severe pain in the symphysis. **B,** healing of these changes after 3 months of restricted activity.

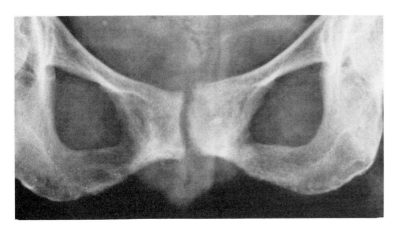

FIG 81.
Symphyseal changes in a 27-year-old woman with pelvic ring instability who has not borne children.

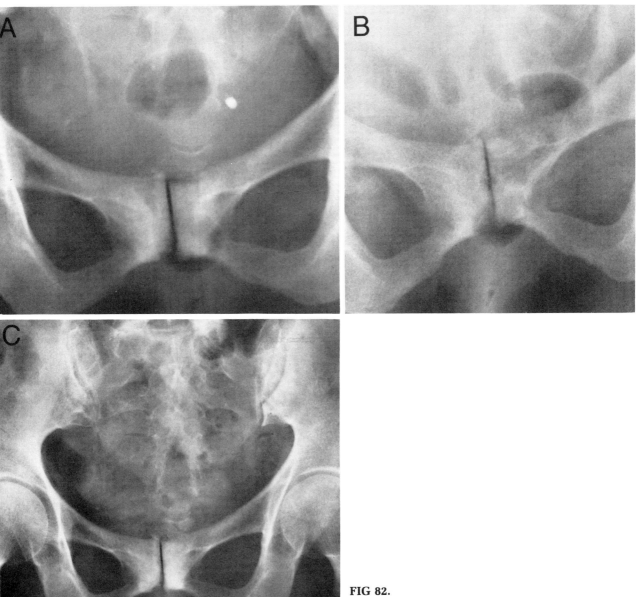

FIG 82.
A, symphyseal changes in a 45-year-old woman with pelvic ring instability. **B** and **C,** standing on one leg and then the other demonstrates laxity of the symphysis pubis.

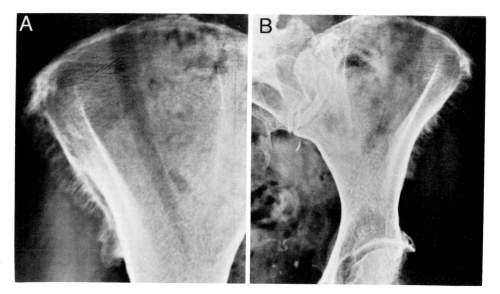

FIG 83.
A and B, "tug lesions" at the margins of the ilia in a 75-year-old woman.

will often demonstrate that a stress injury existed before the sudden onset, indicating that a partial avulsion had become complete.[8]

Radiographically, the healing of these avulsions may simulate the findings of a neoplasm. Therefore, a careful history of athletic activity is most important.

Figure 84 illustrates the apophyses and muscles of the pelvis that are most commonly involved in avulsion injuries.

THE ILIAC CREST

Figure 85 shows avulsion of the lateral aspect of the left iliac crest in a 16-year old boy. The apophysis of the crest may form from several separate foci, and such fragmentation is not necessarily abnormal. However, separation of a portion of the apophysis, as in this case, is diagnostic. Comparison with the opposite side is often useful.

Apophysiolysis of the iliac crest is seen largely in athletes involved in running,[156] jumping,[157] and figure skating[158, 159] and in those whose activities require forceful contraction of the abdominal muscles.[160] Avulsion of the crest of the ilium may also occur in adults after closure of the apophysis as a result of forceful muscle pull (Fig 86).

THE ANTEROSUPERIOR ILIAC SPINE

Avulsion of the anterosuperior iliac spine is an adolescent injury seen in football, baseball, sprinting, bicycle racing, and track.[159] A single case has been reported in a break dancer.[161] The avulsion is the result of traction of the sartorius/tensor fascia lata (Fig 87). The apophysis does not regain its original position with healing but progressively

lays down new bone, creating a gradually enlarging protuberance that may be mistaken for an osteochondroma (Figs 88 and 89). Figure 90 shows an example of old and new avulsions of the anterosuperior iliac spine in the same individual.

THE ANTEROINFERIOR ILIAC SPINE

Avulsions of the anteroinferior iliac spine are seen in soccer and field hockey players[159] and runners.[8] In young children in whom the secondary apophysis has not yet appeared, the pull of the insertion of the rectus femoris muscle produces irregularity of the underlying bone, which might be mistaken for evidence of a malignant neoplasm (Fig 91). After the secondary ossification center appears, the avulsion is easier to detect (Fig 92). When these lesions heal, they leave residual deformity (Fig 93).

ISCHIAL AVULSIONS

The insertion of the powerful hamstring muscles at the ischium produces a variety of radiologic alterations as a product of muscle pull. Figure 94 shows the location of the apophyses in this area that become implicated in these alterations. Avulsive lesions affecting these apophyses are seen primarily in athletes involved in baseball, figure skating, ice hockey,[159] track,[162] and football.[163] Figure 95 shows the typical ischial lesions seen in adolescence. At times these innocent lesions may become very prominent (Fig 96) and lead the unwary physician to perform a biopsy.[163, 164] Chronic stress may produce a sclerotic reaction in the ischial apophysis that might be alarming if the mechanism of such an appearance is not understood (Fig 97).

Total avulsion of the ischial apophysis may occur (Fig

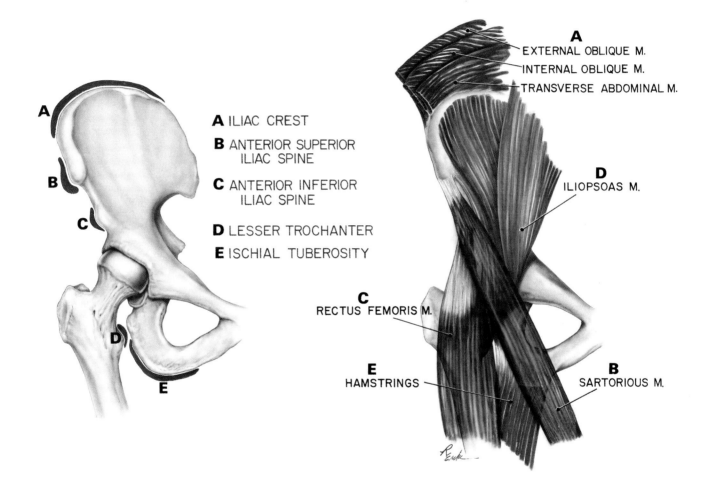

A ILIAC CREST

B ANTERIOR SUPERIOR
ILIAC SPINE

C ANTERIOR INFERIOR
ILIAC SPINE

D LESSER TROCHANTER

E ISCHIAL TUBEROSITY

A
EXTERNAL OBLIQUE M.
INTERNAL OBLIQUE M.
TRANSVERSE ABDOMINAL M.

D
ILIOPSOAS M.

C
RECTUS FEMORIS M.

E
HAMSTRINGS

B
SARTORIOUS M.

FIG 84.
The apophyses and muscles of the pelvis most often implicated in avulsive injuries of the pelvis.

FIG 85.
Avulsion of a portion of the apophysis of the
left iliac crest in a 16-year-old boy (*arrow*).

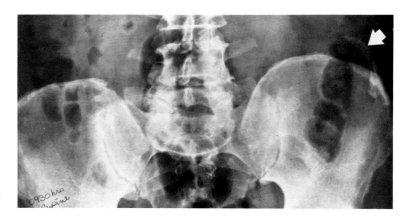

FIG 86.
Avulsion of the crest of the ilium in a 60-year-old
man (*arrow*).

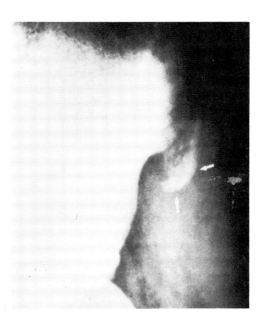

FIG 87.
Avulsion of the anterosuperior iliac spine (*arrows*).

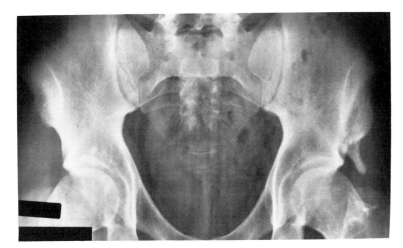

FIG 88.
Old avulsions of the anterosuperior iliac spines.

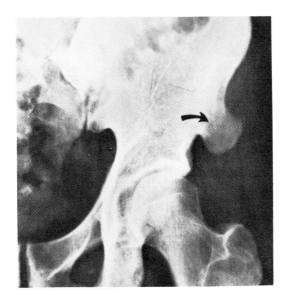

FIG 89.
Old healed avulsion of the anterosuperior iliac spine simulating
an osteochondroma (*arrows*).

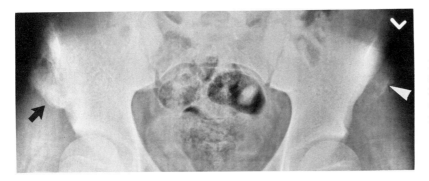

FIG 90.
Old healed avulsion of the anterosuperior
iliac spine on the right (*arrow*), and a new
avulsion on the left (*arrowhead*), in a 13-
year-old football player.

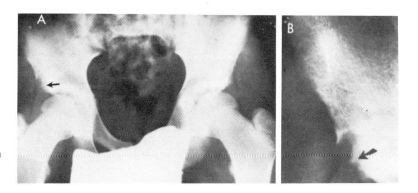

FIG 91.
A, changes of avulsion of the anteroinferior iliac spine before the appearance of the secondary apophysis (*arrow*). **B,** magnification view showing irregularity of underlying bone (*arrow*).

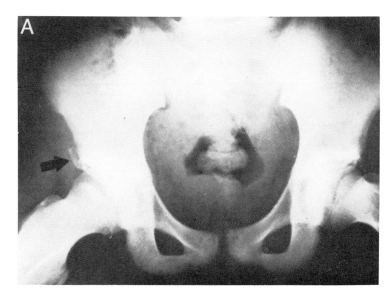

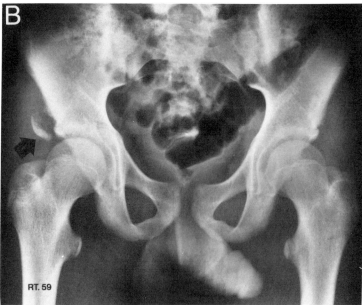

FIG 92.
A and **B,** examples of avulsions of the apophysis of the anteroinferior iliac spine (*arrows*).

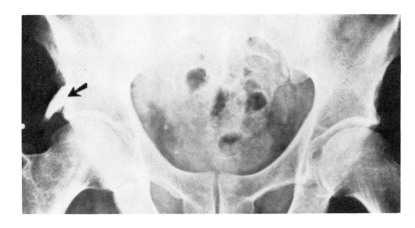

FIG 93.
Old healed avulsion of the anteroinferior iliac spine (*arrow*).

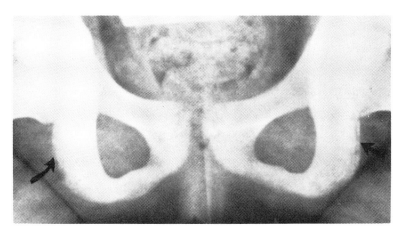

FIG 94.
Pelvis of a 13-year-old showing the locations of the ischial apophyses (*arrows*), best seen on the patient's left side.

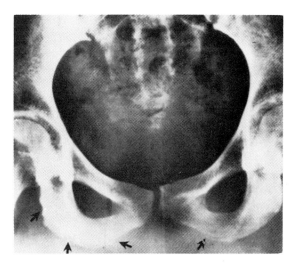

FIG 95.
Tug lesions of the right ischium in an adolescent (*arrows*). The arrow on the patient's left side indicates a normal apophysis.

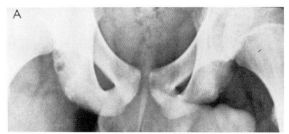
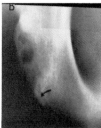

FIG 96.
A, tug lesions of the ischia in an adolescent boy.
B, tomogram of the right ischium. Note the open
apophysis (*arrow*).

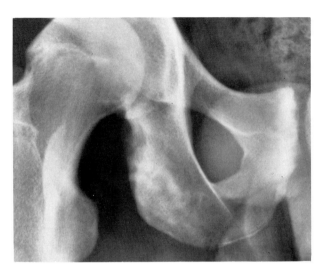

FIG 97.
Chronic ischial tug lesions in an 18-year-old football player
evidenced only by sclerosis.

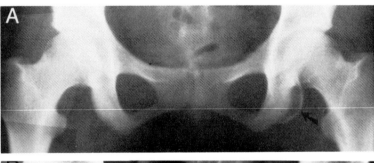

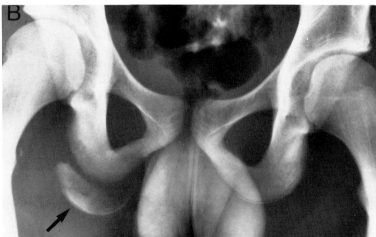

FIG 98.
Avulsion of the ischial apophysis (*arrows*) in a 14-
year-old girl **(A)** and a 16-year-old boy **(B).**

98), and this lesion is at times mistaken for an expanding neoplasm of the ischium (Fig 98,A). When this lesion heals, it leaves a rather characteristic deformity of the ischium that might be mistaken for an osteochondroma (Fig 99).

Avulsion of the ischial apophysis may be associated with a stress reaction at the insertion of the adductor muscles at the femur. In the case illustrated in Figure 100, the avulsed ischial apophysis was missed on the original ex-

amination and attention was directed to the periosteal reaction in the femur, which raised the specter of malignant neoplasia.

STRESS LESIONS OF THE SYMPHYSIS PUBIS

The adductor muscles originating near the symphysis pubis include the adductors brevis and longus and the gracilis. In athletes, chronic stress of these muscles may lead to osseous changes in the symphysis pubis. These muscle attachments are shown in Figure 101. This phenomenon has been termed *osteitis pubis* in the sports medicine and orthopedic literature[165–168]; it is also known as the gracilis syndrome.[162, 169] The term *osteitis pubis* unfortunately leads to confusion with the syndrome that follows bladder, prostate, or uterine surgery or childbirth.[168, 170]

Stress-induced avulsive injury near the symphysis pubis is characterized by gradual onset of localized pain and tenderness of the symphysis and in the adjacent muscles. On physical examination, there is point tenderness in the symphyseal area and adjacent muscles. Forceful adduction of the thighs against resistance is especially painful. The disease has been described in road walking,[165] track,[171] football,[171] and in runners[171, 172] and soccer players.[166, 168]

The radiologic findings are similar to those found in infection and true osteitis pubis and consist of mixed bone lysis, fragmentation, and sclerosis on one or both sides of the symphysis. The changes may extend into the inferior pubic ramus.[171] Figures 102 and 103 show these types of changes in athletes. In addition, there is an association between stress fracture of the pubis and this type of symphyseal avulsive injury in marathon runners[173] (see Fig 78).

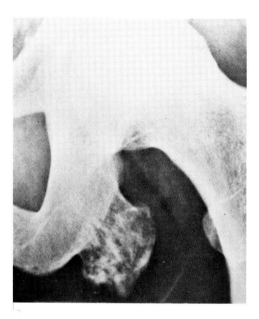

FIG 99.
Healed avulsion of the ischial apophysis simulating an osteochondroma.

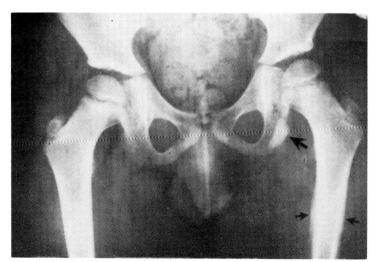

FIG 100.
Avulsion of the ischial apophysis (*large arrow*) associated with stress reaction in the femur at the insertion point of the adductor muscles (*small arrows*).

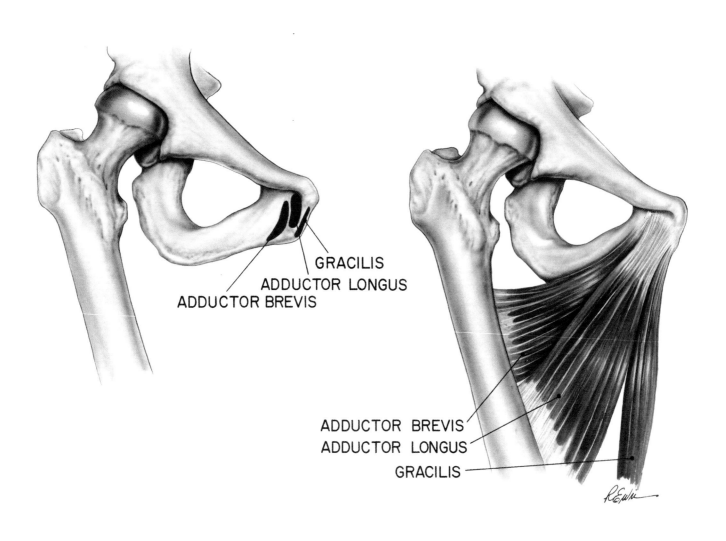

FIG 101.
The muscles and their insertion points about the symphysis pubis.

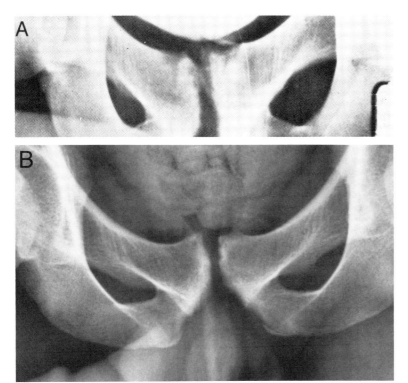

FIG 102.
A, irregularity and sclerosis of bone in the symphysis pubis in a 19-year-old hurdler. **B,** symphyseal irregularity in a 20-year-old soccer player.

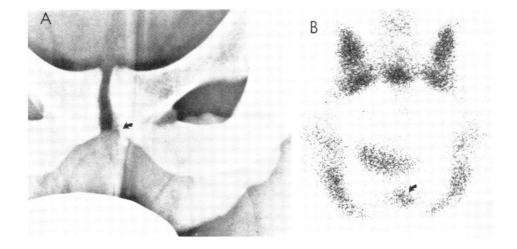

FIG 103.
A, irregularity of the left side of the symphysis pubis in a track athlete (*arrow*). **B,** bone scan showing "hot spot" in the same area (*arrow*).

7

The Lower Extremity

THE FEMUR

The Femoral Head

Stress Fractures

Femoral stress fractures are seen in the femoral neck, the midshaft, and the distal metaphysis. Of these, the femoral neck fractures are three to four times more frequent than the other fractures.[7] Experience with fractures in this location has been derived largely from data concerning military recruits[7, 8, 174–177] and runners.[174, 176, 178–181] Although this fracture is seen in young adults for the most part, it has also been described in children.[182–184]

The usual symptoms consist of pain in the hip on weight bearing and limp of several days' to several weeks' duration. Early presumptive diagnosis is possible by radionuclide imaging, which evidences limited areas of homogeneous subcapital increased uptake in the femoral neck[28, 185, 186] (Fig 104). Roentgenograms become diagnostic between 1 and 3 weeks after the onset of symptoms. Morris and Blickenstaff[7] have divided the radiographic findings into three categories:

1. those with sclerosis and compensatory internal or periosteal callus and generally not extending completely across the neck (Fig 105);
2. those with a fracture line in the calcar region or completely across the neck, but without displacement (Figs 2 and 106);
3. those with displacement.

The fracture line will become increasingly apparent even with bed rest, which is the appropriate treatment unless displacement has occurred. Once displacement has occurred, internal fixation is required. Prompt recognition of these fractures is important to avoid the evolution of complete fracture and displacement.

Although stress fracture of the femoral neck is seen primarily in young adults, it is occasionally seen in children before closure of the epiphyses.[183, 184, 187] Figure 107 illustrates a stress fracture of the left femoral neck in an 11-year-old soccer player who kicked with his right foot, but developed pain in his left hip. Plain films were normal, but tomography showed a stress fracture of the left femoral neck. This was confirmed by nuclear scanning and magnetic resonance. The stress to the boy's left femoral neck was apparently the result of repeated fixation of his left hip while he kicked with his right foot.

Stress Reactions

Round to oval radiolucencies surrounded by a thin zone of sclerosis are often found in the proximal superior quadrant of the adult femoral neck. In my experience, they seem to be present in young people particularly. These lucencies have been identified by Pitt et al.[188] as cavities formed by herniation of synovium through defects in the surface of the reactive area of the femoral neck and have been termed "herniation pits" (Fig 108). It has been proposed that the reactive area results from mechanical, abrasive effects of the adjacent overlying hip capsule and the head of the rectus femoris muscle and the more laterally placed iliopsoas muscle. Figure 109 shows bilateral herniation pits in a 25-year-old man, which are demonstrable by computed tomography (CT).

These findings are generally considered to be clinically unimportant. There is, however, at least one report suggesting that removal of this or a similar lesion resulted in relief of hip pain.[189]

Osteochondrosis Dissecans

Isolated osteochondrosis dissecans of the femoral head may be seen in young adults in the absence of trauma or familial occurrence. However, it does appear to be related to musculoskeletal stress in some individuals and has been described in soldiers and athletes[190, 191] (Fig 110).

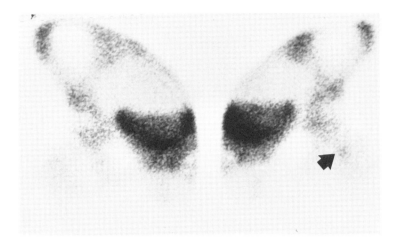

FIG 104.
Nuclear scan of a 30-year-old runner with left hip pain showing increased uptake in the left femoral neck at the site of a stress fracture (*arrow*). The plain films were normal at the time.

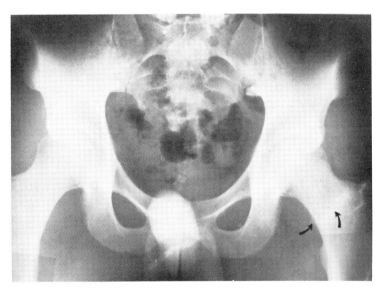

FIG 105.
Subtle stress fracture of the femoral neck in a 20-year-old soldier seen as a faint band of osteosclerosis and a small puff of new bone at the inferior aspect of the femoral neck (*arrows*).

The Femoral Shaft

Stress Fractures

Stress fractures of the shaft of the femur may be seen in the subtrochanteric area,[192] the central shaft,[193-196] the distal metaphysis, or even through the distal femoral epiphysis.[197] The original descriptions of these fractures were derived largely from military life,[7, 8, 194] but they are now seen most commonly in athletes, particularly runners.[192, 193, 195, 196]

The radiographic signs, like those seen in other stress fractures, may range from early periostitis to frank fractures with displacement (Fig 111). The principal problem encountered in these lesions is the misinterpretation of the reactive changes as evidence of malignant neoplasm, particularly osteosarcoma (Fig 112). This is a significant problem in stress fractures in children, because neoplasm, rather than stress injury, is apt to come to mind initially.

In children stress fractures may also occur in the distal femoral epiphysis and are evidenced by widening of the epiphyseal plate[197] (Fig 113).

Stress Reactions

Early stress reactions in the femoral shaft also raise suspicion of malignant neoplasm owing to the appearance of new bone formations along the shaft (Fig 114). As these reactions consolidate and heal, they are less ominous in their radiographic appearances (Fig 115).

A remarkable case of stress reaction in the femur and calcaneus in a 20-year-old break dancer has been described by Ihmeidan et al.[198] and is illustrated in Figure 116. Ballmer and Bessler[199] reported another remarkable case of a stress reaction of the anterior femur simulating an osteochondroma in a Swiss cheese maker who lifted heavy cheeses from a shelf onto his right thigh (Fig 117).

Another rather typical area of stress reaction in the femur is in the region of insertion of the adductor brevis

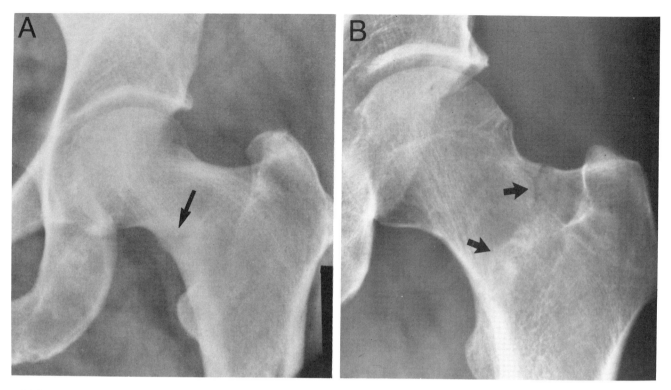

FIG 106.
A and **B,** two examples of stress fracture of the femoral neck showing incomplete lucency and sclerosis of portions of the femoral necks (*arrows*).

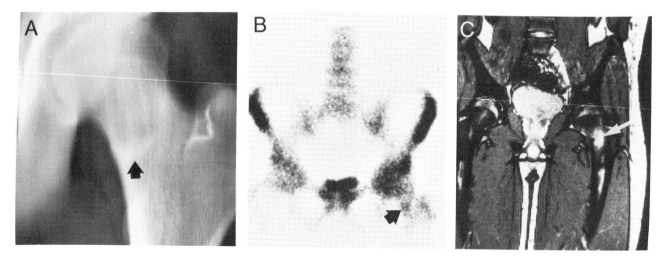

FIG 107.
Stress fracture of the left femoral neck in an 11-year-old soccer player (*arrows*). **A,** laminagram. **B,** nuclear scan. **C,** magnetic resonance scan (TR 2.0 seconds; TE 34 msec).

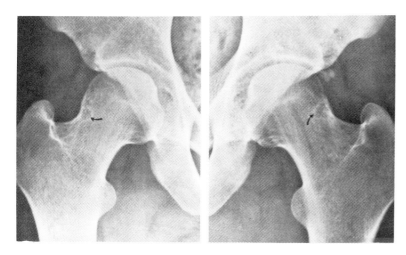

FIG 108.
Bilateral femoral herniation pits (*arrows*) in a 20-year-old female runner with hip pain (**A** and **B**).

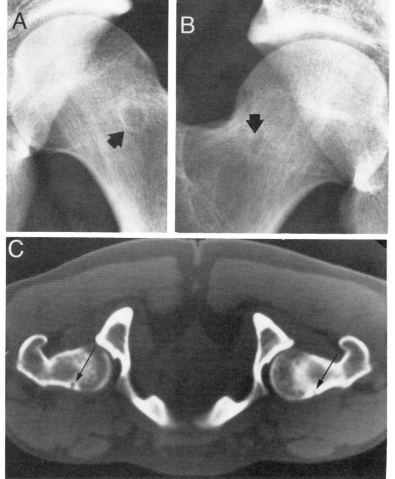

FIG 109.
A and **B,** bilateral herniation pits (*arrows*) in a 25-year-old man, which are also seen on computed tomography (**C**).

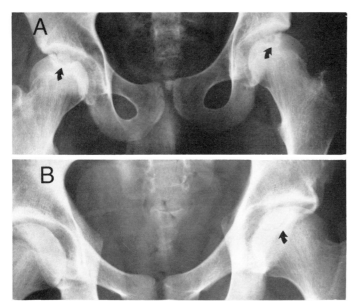

FIG 110.
Two examples of osteochondrosis dissecans of the femoral heads in young adults (*arrows*). **A,** bilateral disease. **B,** unilateral disease.

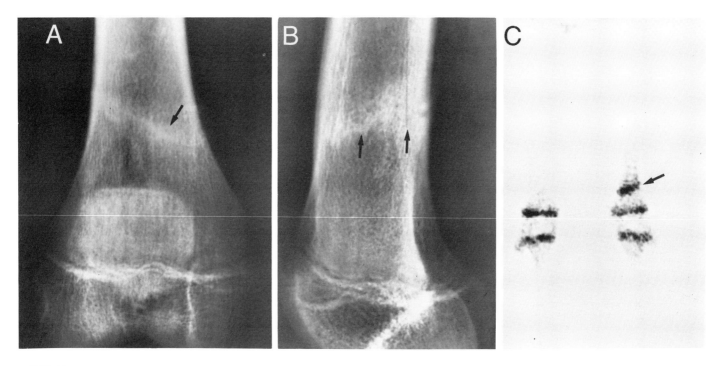

FIG 111.
Typical stress fracture of the left femur in an 11-year-old boy (*arrows*). **A** and **B,** plain film. **C,** nuclear scan.

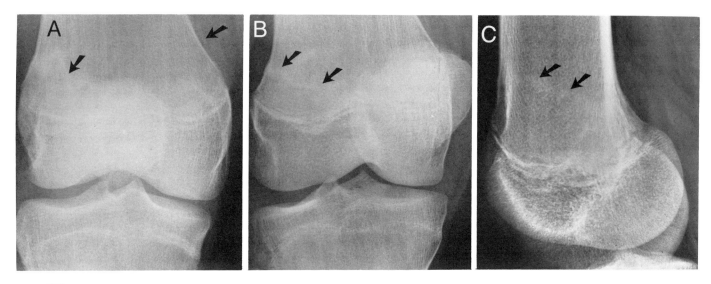

FIG 112.
A to C, stress fracture in the distal femoral metaphysis in a 12-year-old boy originally suspected of having a malignant neoplasm. Note periosteal reaction at medial side (*arrows*).

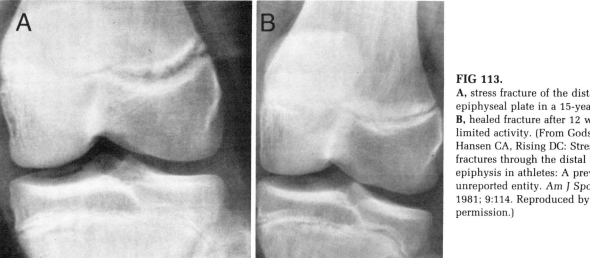

FIG 113.
A, stress fracture of the distal femoral epiphyseal plate in a 15-year-old boy. B, healed fracture after 12 weeks of limited activity. (From Godshall RW, Hansen CA, Rising DC: Stress fractures through the distal femoral epiphysis in athletes: A previously unreported entity. *Am J Sports Med* 1981; 9:114. Reproduced by permission.)

muscle at the cortex of the medial aspect of the femur below the lesser trochanter (Fig 118). Reactions of this type are seen in runners[148] and have also been described in female military basic trainees.[148] Some of these injuries are seen in association with avulsive lesions in the ischium (see Fig 99). These areas of reactivity may be seen on bone scans without plain film findings.[200]

An interesting variation of this lesion is shown in Figure 119, which is similar to that seen in athletes, but this case occured in a 20-year-old woman with Huntington's chorea who was in incessant movement.

Avulsive Lesions of the Femur

The Lesser Trochanter

Avulsion of the secondary center of ossification of the lesser trochanter of the femur is caused by vigorous contraction of the iliopsoas muscle (Fig 120). This lesion is most commonly seen in adolescent athletes but may also occur in youngsters with spastic neurologic disorders. Pain, often with little external evidence of trauma, is the most frequent presenting symptom.[158]

With conservative treatment the symptoms disappear,

FIG 114.
Stress reactions in the femur in a 15-year-old athlete with no history of trauma.

but, if the ossification center is significantly separated from the parent bone, it will not reunite but will remain as a separate ossicle throughout life (Fig 121). Occasionally a bony bridge will develop between the avulsed apophysis and the femur (Fig 122). One may also see tug lesions at the insertion of the iliopsoas muscles reflected as small spurs arising from the lesser trochanter (Fig 123). Another lesion secondary to muscle pull may be seen as separation of the cephalad end of the greater trochanter secondary to action of the gluteus medius and the piriformis muscles (Fig 124).

The Avulsive Posterior Distal Femoral Cortical Defect

Two cortical irregularities are seen in the distal femoral cortex. One, the posteromedial cortical irregularity, which can be seen in the frontal projection in children and adolescents, is, in my opinion, a developmental lesion that is not stress related since it can be seen in other metaphyseal areas in children[201] (Fig 125). This opinion is not shared by all writers on the subject.[202, 203] However, there does appear to be universal agreement that the other cortical lesion, the posterior femoral excavation, is secondary to musculoskeletal stress.[204]

The latter lesion is seen as an excavation in the medial side of the posterior aspect of the distal femur and may present as an oval or circular radiolucency with surrounding sclerosis in the frontal projection and as an excavated cortical irregularity in the lateral projection (Fig 126). The

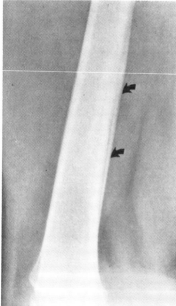

FIG 115.
A and **B,** stress reaction of the femur in a 16-year-old female runner (*arrows*).

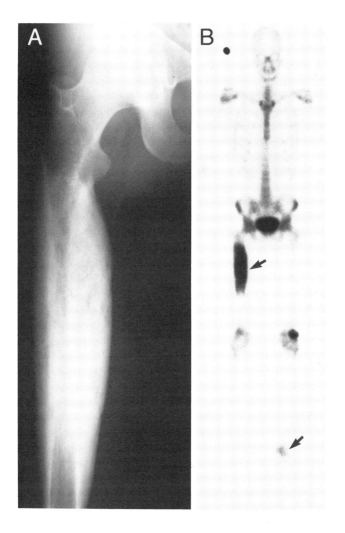

FIG 116.
A and **B,** striking stress reaction in the femur and calcaneus (*arrows*) in a 20-year-old break dancer. (From Ihmeidan IH, Tehranzadeh J, Oldham SA, et al: Case report 443. *Skeletal Radiol* 1987; 16:581. Reproduced by permission.)

location of the defect at the osseous site of attachment of the medial head of the gastrocnemius muscle supports a stress-related pathogenesis. These cortical lesions in the distal femur are not "hot" on bone scanning, a characteristic that helps differentiate them from other diseases.[205] This lesion also carries the misnomer of the desmoid tumor or desmoid lesion, which serves only to confuse its nature.

When the posterior excavation heals, a persistent fossa is often left behind and remains throughout life (Fig 127,A). At times some residual cortical irregularity remains (Fig 127,B).

The significance of both these distal femoral cortical defects is only that they may simulate malignant neoplasm. If one is aware of their characteristic appearance and nature, biopsy should not be necessary.[206]

Tug Lesions of the Distal Femur

The insertion of the distal ends of the abductors of the thigh, when stressed, may lead to a tug lesion of the cortex of the distal femoral shaft that is at times misinterpreted as an osteochondroma and surgically removed (Fig 128).

These lesions are not necessarily symmetric and can be quite large (Fig 129). Occasionally, a similar lesion may be seen on the lateral side at the site of insertion of the abductors (Fig 130) or at the site of insertion of the lateral head of the gastrocnemius (Fig 131).

These tug lesions can be distinguished from an osteochondroma by the fact that the cortex beneath them is intact, in contrast to the osteochondroma, in which the trabeculae of the shaft sweep directly into the lesion without intervening cortex.

Pelligrini-Stieda Disease

Pelligrini-Stieda disease is regarded as a calcification and eventual ossification of the medial collateral ligament due to avulsion or direct trauma.[207] It presents as a painful swelling and is seen radiographically as a calcific density at the medial condyle. Eventually, the painful inflammation subsides with partial or complete resorption of the calcium salts, or the mass becomes ossified and occasionally connected by a pedicle to the femoral condyle. The bony mass constitutes a permanent obstacle to gliding of

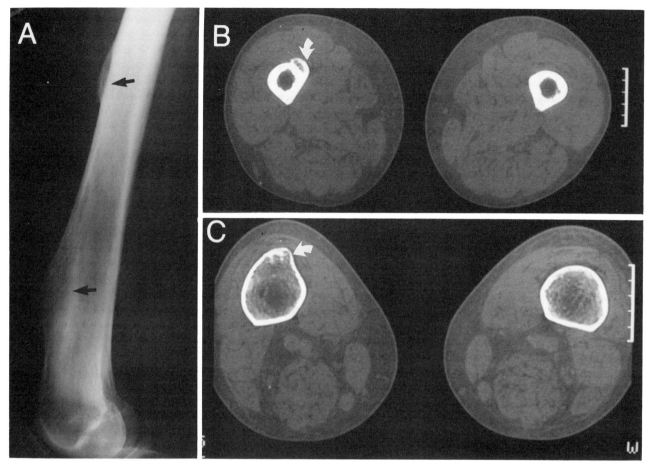

FIG 117.
A, a lateral roentgenogram of the right femur shows a large flat osteochondroma-like exostosis in the distal third of the anterior surface of the femur and a flat, ossified, subperiosteal mass in the middle third of the femoral shaft *(arrows)*. **B,** a computed tomogram through the mid-third of both femora shows a ringlike, ossified, superiosteal mass adjacent to the anterior cortex of the right femur *(arrow)*. The central lucency in the center of the bony protuberance is consistent with old hemorrhage. **C,** a computed tomogram through the distal third of the femora shows a bony protuberance on the anterior aspect of the right femur with a convex bulge of the thinned cortex *(arrow)*. Superficial densities are noted in the underlying trabecular marrow. (From Ballmer PE, Bessler WT: Case report 495. *Skeletal Radiol* 1988; 17:382. Reproduced by permission.)

the ligament and may restrict movement of the knee[208] (Fig 132). Rarely a similar process may be seen at the lateral condyle (Fig 133). Experience has shown that the calcification becomes radiographically apparent about 3 weeks after the onset of symptoms. Three-phase bone scanning antedates the radiographic changes and shows a focal area of increased flow at the medial aspect of the knee.[208] The calcification progresses and stabilizes over the next few months. It rarely is a cause of permanent disability.[207]

Osteochondrosis

Osteochondrosis dissecans of the femoral condyles is a common adolescent lesion that may occur spontaneously or as part of a familial disorder.[209–211] It appears to occur with an inordinate incidence in young athletes and there-

fore may indeed be stress induced.[190, 212] It is most common in the lateral aspect of the medial femoral condyle (Fig 134, A and B) but may also be seen on the anterior articular surfaces of the condyle (Fig 134, C and D). Rarely, it is seen in the posterior femoral condyles.[213, 214] The disease may be unilateral or bilateral. The osseous cartilage fragment may remain intact at its point of origin or form a loose body in the joint.

Joint scintigraphy in these patients shows a pattern typical of fractures elsewhere in the skeleton with increased activity surrounding the lesion.[215]

The degree of Tc-99m pyrophosphate uptake is a measure of both osteoblastic activity and regional blood flow. The level of activity in a symptomatic knee will indicate the remaining potential for healing of the osteochondritic fragment. The scans with high activity indicate repair po-

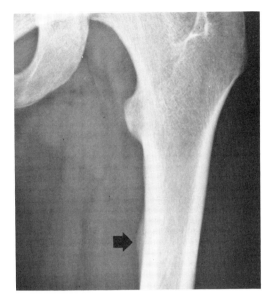

FIG 118.
Stress reaction at the insertion of the adductor brevis in a 13-year-old girl (*arrow*).

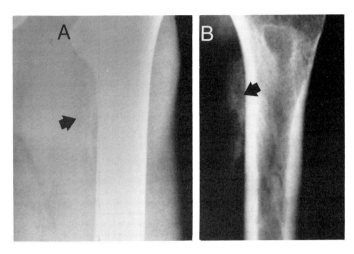

FIG 119.
A and **B,** adductor brevis stress reaction (*arrows*) in a woman who is hyperactive secondary to Huntington's chorea.

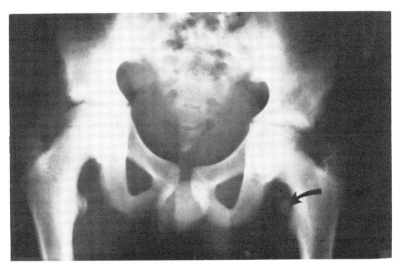

FIG 120.
Avulsion of the apophysis of the lesser trochanter (*arrow*).

tential. Conversely, those that show a low level of activity have a poorer prognosis for spontaneous healing.[209]

A similar and rather typical lucent articular lesion has been described in the anterior aspect of the lateral femoral condyle in athletes.[216, 217] The lesion occurs on the lateral femoral condyle just distal to the epiphyseal plate at the level of the patella. These patients have patellar femoral pain and symptoms consistent with "chondromalacia patellae" or one of the patellar pain syndromes. The lesion is assumed to be similar to osteochondrosis dissecans and is "hot" on radionuclide bone scan. We have seen the identical process in an 18-year-old equestrienne (Fig 135) and in a 17-year-old athlete (Fig 136). The lesion is best seen in the lateral projection. A sclerotic lesion that I believe represents the healed stage of this disease is shown in Figure 137.

Friction by the Pes Anserinus

A syndrome related to friction by the pes anserinus on the medial femoral condyle has been reported by Fornasier et al.[218] A 26-year-old man had a 1-year history of

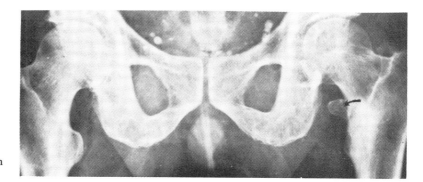

FIG 121.
Old ununited apophysis of the lesser trochanter in a 60-year-old man (*arrow*). Note the abnormal configuration of the normal site of the apophysis in the femur.

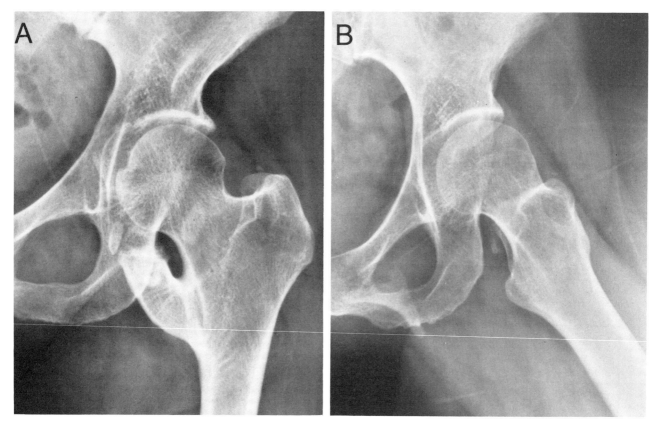

FIG 122.
A and **B,** old avulsion of the apophysis of the lesser trochanter in a 30-year-old patient with spastic cerebral palsy showing the development of a bony bridge between the femur and the avulsed apophysis.

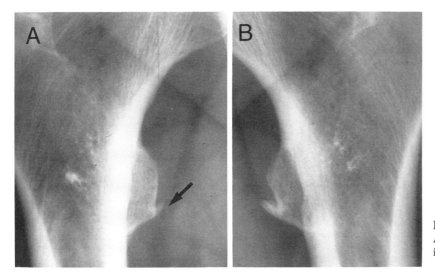

FIG 123.
A and **B**, tug lesions at the insertion of the iliopsoas muscles (*arrow*).

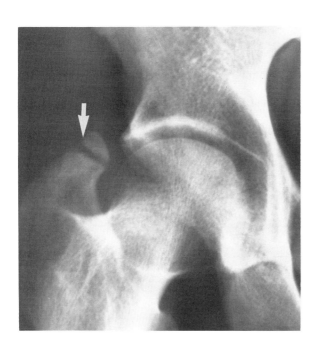

FIG 124.
Separations of the tip of the greater trochanter secondary to muscle pull (*arrow*).

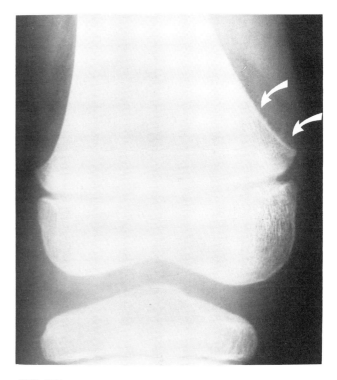

FIG 125.
The medial distal femoral cortical defect (*arrows*), a lesion that is probably not stress related.

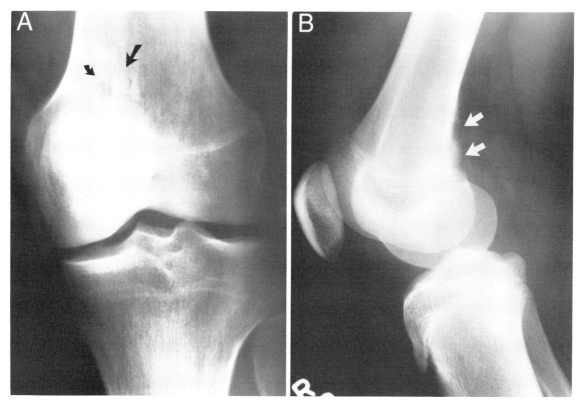

FIG 126.
A and **B**, typical posterior femoral cortical excavation (*arrows*).

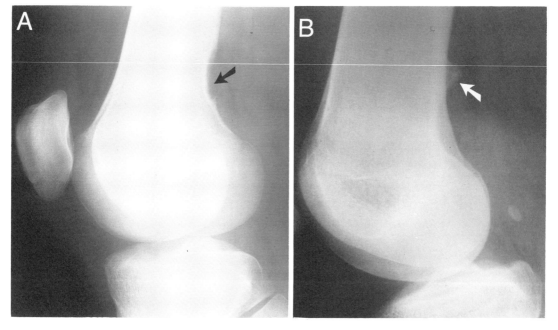

FIG 127.
A, residual deformity of the posterior femoral cortex (*arrow*) resulting from the posterior avulsive lesion in a 30-year-old man. **B**, 18-year-old man with residual deformity of the posterior femoral cortex and some cortical irregularity superiorly.

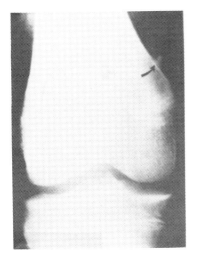
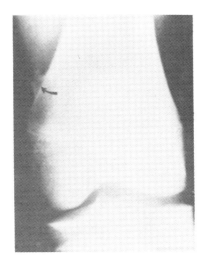

FIG 128.
A and **B,** bilateral femoral tug lesions in a 12-year-old girl (*arrows*).

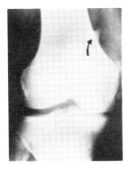

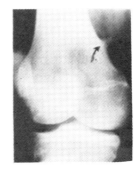

FIG 129.
A to **C,** large femoral tug lesion in a 19-year-old football player (*arrows*).

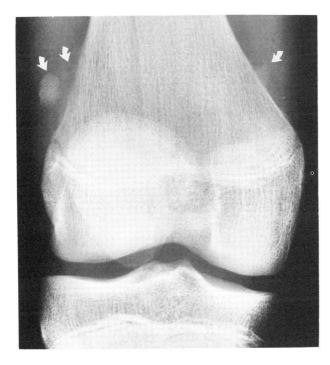

FIG 130.
Bilateral tug lesions of the femur (*arrow*).

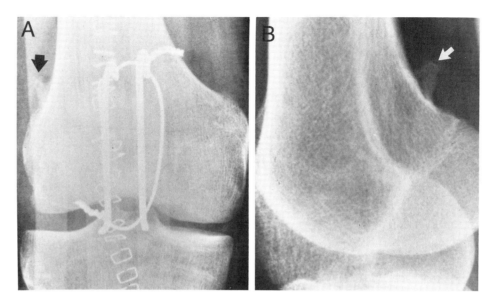

FIG 131.
A and **B,** tug lesion probably at insertion of the lateral head of the gastrocnemius (*arrows*).

FIG 132.
Long-standing Pellegrini-Stieda disease with ossification of the prior calcific deposit (*arrows*). **A,** anteroposterior view. **B,** tunnel projection.

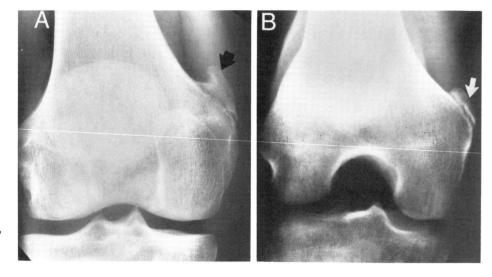

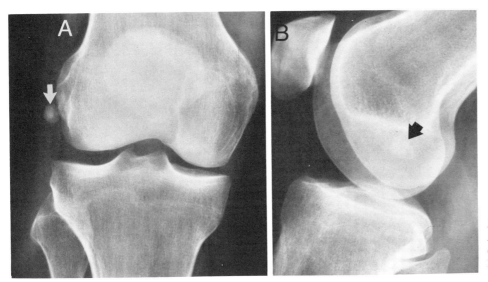

FIG 133.
A and **B,** calcification in the lateral collateral ligament (*arrows*; reverse Pellegrini-Stieda disease).

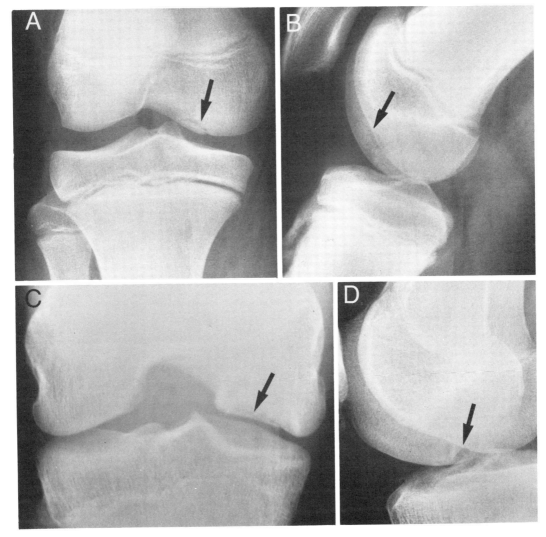

FIG 134.
A and **B,** osteochondrosis dissecans of the femur in its classic location in the lateral aspect of the medial femoral condyle (*arrows*) in an 11-year-old boy. **C** and **D,** osteochondrosis dissecans of the medial femoral condyle with involvement of the articular surface (*arrows*) in a 16-year-old boy.

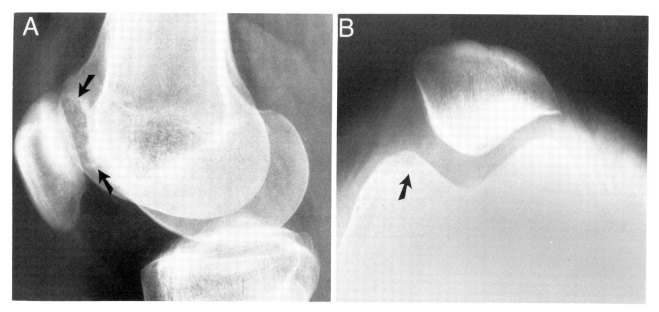

FIG 135.
A and **B,** presumed osteochondrosis dissecans in the lateral femoral condyle in an 18-year-old equestrienne (*arrows*).

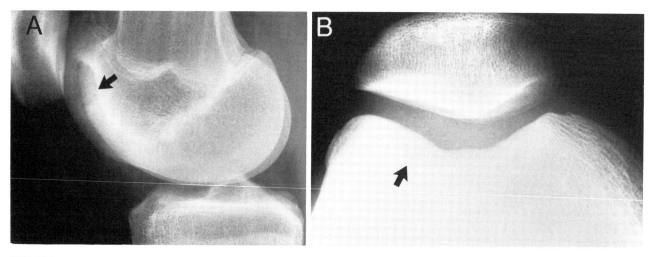

FIG 136.
Osteochondritic lesion of the lateral femoral condyle in a 17-year-old male athlete (*arrows*). **A,** lateral projection. **B,** tangential projection.

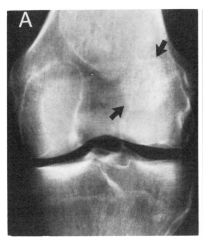
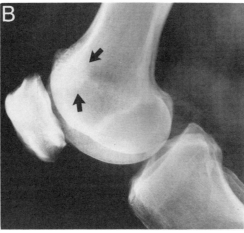

FIG 137.
A and **B,** sclerotic lesion in the lateral femoral condyle (*arrows*) in a 38-year-old woman with a history of knee pain. Note similarity of the location of the lesion in Figure 128 and the degenerative changes in the patellofemoral compartment.

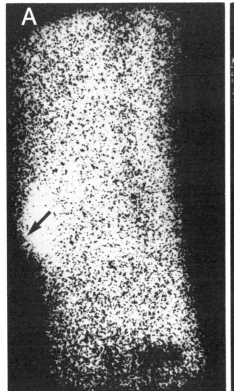
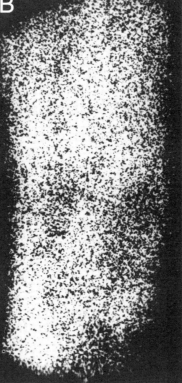

FIG 138.
Increased uptake in the medial femoral condyle of the left knee due to friction by the pes anserinus tendon **(A)** *(arrow).* The right knee is shown for comparison **(B).** (From Fornasier VL, Czitrom AA, Evans JA, et al: Case report 398. *Skeletal Radiol* 1987; 16:57. Reproduced by permission.)

pain on the medial aspect of the left knee. Plain films of the knee were normal, but a ^{99m}Tc bone scan demonstrated an area of increased uptake at the medial condyle (Fig 138). Excision confirmed the reactive nature of the lesion, which had initiated a periosteal proliferation. Other syndromes related to friction by musculotendinous or fascial structures have been reported.[219–222]

THE PATELLA

Stress Fractures

Stress fracture of the patella is unusual and seems to have a particular predilection for the adolescent athlete engaged in running, Japanese fencing, football, soccer, and long-distance walking. The fracture may be horizontal[8, 223]

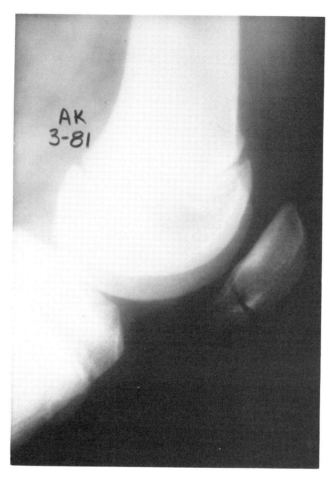

FIG 139.
Horizontal stress fracture of the patella in a 12-year-old boy. (From Dickason JM, Fox JM: Fracture of the patella due to overuse syndrome in a child: A case report. *Am J Sports Med* 1982; 10:248. Reproduced by permission.)

or longitudinal[224] (Figs 139 and 140). The fracture is usually well seen in tangential (skyline) projection. The transverse fracture is due to muscular traction stresses; the vertical fracture is secondary to the forces that compress the patella against the lateral femoral condyle. Many authors suggest that the pull of the quadriceps makes the lateral facet incongruent with the lateral femoral condyle from 20° to 50° of flexion and yields more compressive force in the lateral than the medial facet. The predominance of pathologic lesions on the lateral side of the joint, as seen in dislocations, subluxations, and the lateral pressure syndrome, supports the theory of incongruity between the lateral facet and the lateral condyle.[224] This concept is also discussed below in the section dealing with the dorsal patellar defect.

Avulsions

Avulsion fractures of the body of the patella are radiographically identical to those that are acquired as the result of direct trauma. The history of the sudden onset of pain, at times with an audible sound, is diagnostic of the avulsion fracture[225] (Fig 141).

Jumper's knee is a syndrome seen in athletes who are involved in sports activities that place repetitive stresses on the extensor mechanism of the knee, particularly basketball. The disease is the result of repetitive microtrauma concentrated at either the superior or inferior poles of the patella. The end stage of the disease is a complete rupture of the tendinous attachment to the involved pole. The symptoms of pain and tenderness at one of the poles of the patella are associated with ossicles in the quadriceps or patellar tendon, elongation of the poles of the patella, ossicles near the tibial tubercle, and toothlike projections on the tangential view of the patella (tooth sign).[226] Fragmentation of the lower pole of the patella has also been described in cerebrospastic children as a result of abnormal

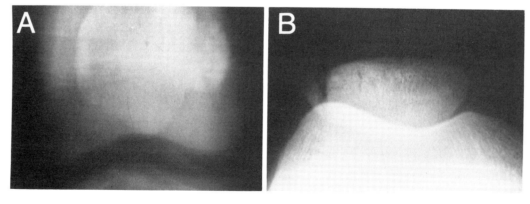

FIG 140.
A and **B,** lateral longitudinal stress fracture of the patella in an 11-year-old boy. (From Iwaya T, Takatori Y: Lateral longitudinal stress fracture of the patella: Report of three cases. *J Pediatr Orthop* 1985; 5:73. Reproduced by permission.)

crepitation, and inability to extend the knee. In rare cases, the rupture may occur in the main substance of the tendon. Precise diagnosis is usually made radiographically. The knee should be filmed in lateral projection in flexion. Rupture of the tendon results in high position of the patella, obliteration of the outline of the tendon, and disruption of the infrapatellar fat pad. Occasionally, a small bone fragment from the inferior aspect of the patella or from the tibial tubercle may be visualized[232] (Fig 146).

Tendon rupture can be imaged by magnetic resonance, which can produce additional information such as the location of the tear, the condition of the tendon, and the appearance of the surrounding soft tissues. Magnetic resonance can also demonstrate incomplete tendon tears[231] (Fig 147).

The Dorsal Defect of the Patella

The dorsal defect of the patella is a poorly understood lesion that may be stress related. This entity is seen as a round, lytic lesion with well-defined margins, located in the superolateral aspect of the patella adjacent to the subchondral bone. Arthrography indicates that the articular

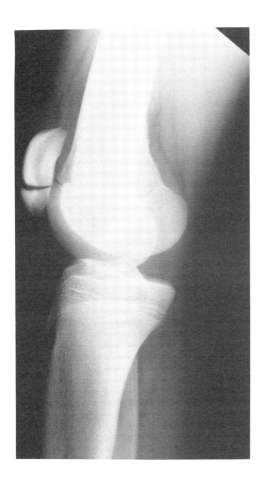

FIG 141.
Avulsion fracture of the body of the patella in a 13-year-old boy, which occurred during a high jump. (From Hughes AW: Case report: Avulsion fracture involving the body of the patella. *Br J Sports Med* 1985; 19:119. Reproduced by permission.)

stresses. This will eventually result in an elongated patella.[227, 228] Figures 142 and 143 illustrate the typical changes of jumper's knee in athletes. Figure 144 shows an advanced stage of the disease with ossicles in the patellar tendon as well as at the inferior pole of the patella. Figure 145 shows ossification of the insertion of the quadriceps and patellar tendons seen in older individuals as a result of musculotendinous stress.

Spontaneous complete rupture of the patellar tendon is rare. It has been described in weight lifters[229] and as a result of twisting injury.[230] For the most part these ruptures occur in association with systemic disease, such as systemic lupus erythematosus, renal failure, secondary hyperparathyroidism, and rheumatoid arthritis,[226, 231, 232] or after systemic steroid ingestion[232] or local steroid injections.[233, 234] These reports suggest that the tendons are weakened by systemic disease or by steroids.

The disruption of the tendon usually occurs at the point of insertion at the tibial tubercle and results in pain,

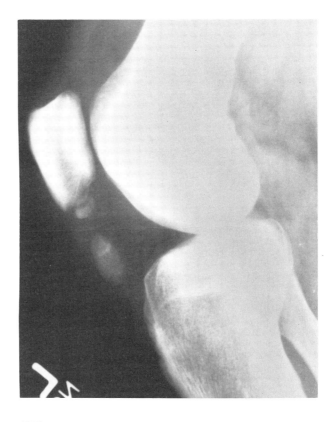

FIG 142.
Advanced changes of jumper's knee in a 28-year-old hurdler with ossification in the patellar tendon.

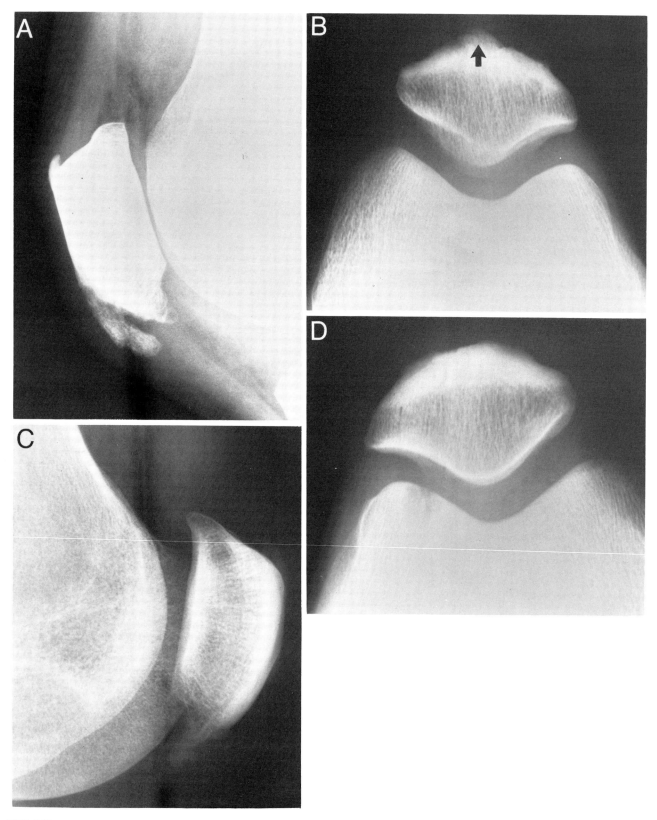

FIG 143.
Jumper's knees in a 33-year-old basketball player, showing ossicles at tendon insertions and tooth sign in the right knee (*arrow*). **A** and **B**, right knee. **C** and **D**, left knee.

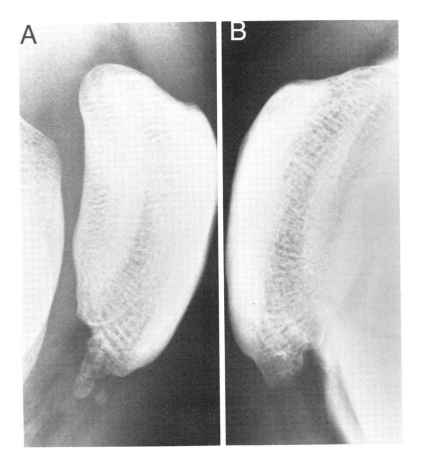

FIG 144.
A and **B,** jumper's knees in a 20-year-old basketball player. Note ossicles in patellar tendon and elongation of the patellae.

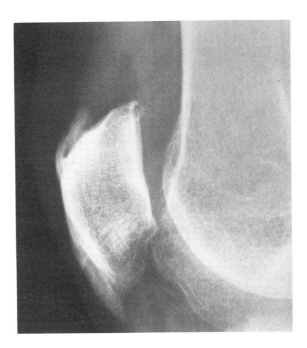

FIG 145.
Ossification at the insertions of the quadriceps and patellar tendons in a 60-year-old athletic man.

cartilage overlying it is intact. The dorsal defect is found in either sex and may be bilateral[235, 236] (Fig 148). The lesion is seen in adolescents and young adults and usually heals spontaneously since it is not seen in middle-aged or elderly individuals. In many cases, it is seen as an incidental finding in individuals who are being examined radiographically for trauma, but some patients complain of knee discomfort. It is at times mistaken for osteochondrosis, which typically involves the articular surface.

Recently, work by van Holsbeeck et al.[237] has suggested a relationship between the dorsal patellar defect and the multipartite patella. Their observations provide evidence that the dorsal patellar defect is a stress-induced anomaly of ossification. They describe a patient with a dorsal patellar defect on one side and a partitioned patella on the other (Fig 149) and suggest that both lesions are the result of traction at the insertion of the vastus lateralis muscle, possibly related to dysfunction of the quadriceps mechanism and patellar subluxation. The association of these two lesions may be significant clinically since it is recognized that the partitioned patella may be painful and subject to stress. It has been suggested that the bipartite patella may actually represent a chronic stress fracture[238] (Fig 150).

FIG 146.
A, normal right knee. Normal infrapatellar fat pad *(arrow)*, quadriceps tendon *(Q)*, and patellar tendon *(P)*. **B,** left knee after patellar tendon rupture. High position of patella and disruption of infrapatellar fat pad *(arrow)*. (From Kricun R, Kricun ME, Arangio GA, et al: Patellar tendon rupture with underlying systemic disease. *AJR* 1980; 135:803. Reproduced by permission.)

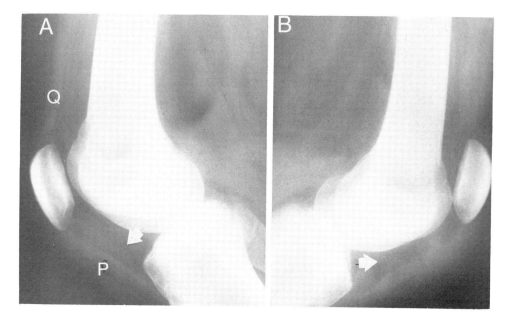

FIG 147.
A, sagittal magnetic resonance image (TR 2,000 msec; TE 60 msec) of the right knee shows complete disruption of the patellar tendon just inferior to its patellar attachment and high-riding patella. The presence of a small fascial remnant *(small arrow)* separated from the remainder of the tendon *(large arrow)* was confirmed at surgery. **B,** left knee shows a similar disruption of the patellar tendon *(arrows)*. (From Gould ES, Taylor S, Naidich JB, et al: MR appearance of bilateral, spontaneous patellar tendon rupture in systemic lupus erythematosus. *J Comput Assist Tomogr* 1987; 11:1096. Reproduced by permission).

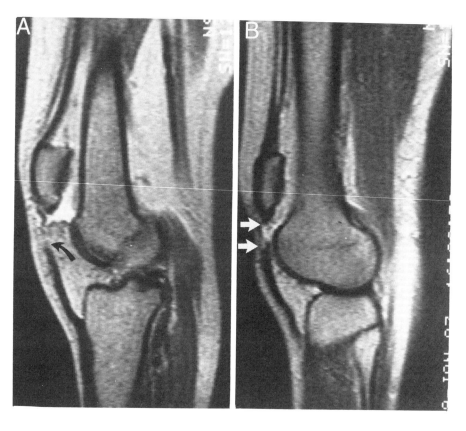

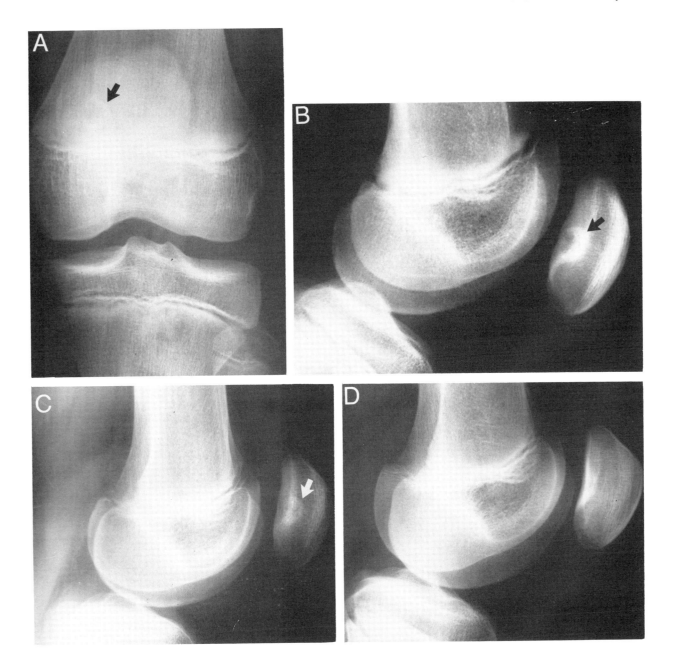

FIG 148.
A and **B,** the dorsal patellar defect in a 13-year-old boy (*arrows*). **C** and **D,** views 1 year later and 18 months later show sclerosis of healing.

Osteochondrosis Patellae

Osteochondrosis dissecans of the patella is relatively rare and may bear a relationship to musculoskeletal stress in adolescent athletes[190, 239] (Fig 151). In patients with patellar subluxation, flexion of the knee will result in repetitive shearing stress on the patellar surface. This is thought to be an important etiologic factor.[240]

The signs and symptoms are nonspecific.[241] The lateral projection of the knee shows the lesions to best advantage and demonstrates a defect in the continuity of the articular surface of the patella with osteochondral fragments lying in a fossa in the bone (Fig 152). In some cases the fragments may be quite large or multiple and may be wholly or partially detached from the articular surface. This entity should not be confused with the dorsal patellar defect described above.

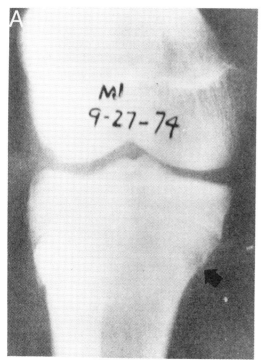

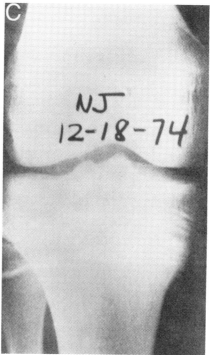

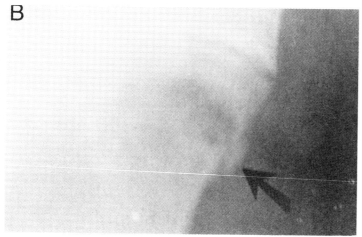

FIG 154.
A, oblique view of the proximal tibia showing irregularity of the cortex below the epiphyseal plate (*arrow*). B, magnification shows a triangular metaphyseal fragment (*arrow*). C, film made 4 months later shows healing with periosteal new bone. There is a suggestion of early closure of the epiphyseal plate. (From Cahill BR: Stress fracture of the proximal tibial epiphysis: A case report. *Am J Sports Med* 1977; 5:186. Reproduced by permission.)

The Proximal Tibial Epiphysis

Stress fracture through the proximal tibial epiphysis is very rare, only one case being found in a 15 1/2-year-old boy who ran 40 to 50 miles per week for 2 months. Radiographic examination showed a Salter-Harris type II fracture of the medial aspect of the proximal tibial epiphysis[243] (Fig 154).

The Medial Tibial Plateau

This fracture is localized 1 to 2 cm below the medial articular surface. It is characterized by pain localized to the anteromedial aspect of the proximal part of the tibia below the joint lines. Weight bearing precipitates the pain and rest relieves it. There is sharply localized tenderness to palpation in the area. Fractures of this type are seen primarily in military recruits.[7] These fractures are usually of the condensing type with the gradual development of a hazy plate-like area of endosteal callus or sclerosis in a line 2 to 3 mm wide across the medial aspect of the metaphysis (Fig 155). During the 2 weeks following the injury, this density increases and slight subperiosteal callus may be present on the medial and posterior aspects of the metaphysis. Rarely, the zone of sclerosis will extend completely across the metaphysis, or a fine fracture line may be present. Displacement does not occur.[7, 8, 244] Early diagnosis requires the use of scans since the radiographic findings are delayed for several weeks.[245]

An unusual stress fracture of the proximal tibial metaphysis extending into the upper articular surface has been described in an 18-year-old runner[246] (Fig 156).

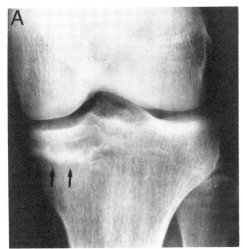
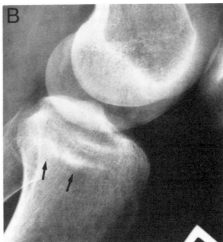

FIG 155.
A and **B**, typical stress fracture of the medial tibial plateau in a 19-year-old soldier (*arrows*).

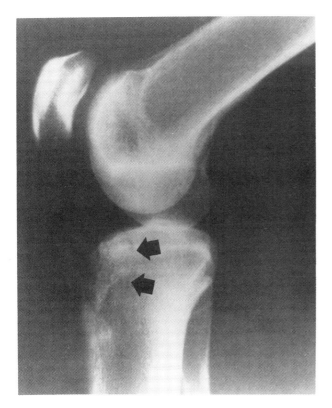

FIG 156.
Unusual stress fracture of the proximal tibial metaphysis extending into the articular surface. (From Trimmings NP: An unusual stress fracture of the upper end of the tibia. *Injury* 1985; 16:348. Reproduced by permission.)

Oblique Fracture of the Posteromedial Tibial Cortex

These are stress fractures developing along the popliteal-soleal line on the posteromedial surface of the tibia and represent the classic runner's stress fracture.[162, 247, 248] These fractures are associated with pain in the proximal portion of the leg with soft-tissue swelling. Early diagnosis is suggested by nuclear scanning, which shows increased uptake in the posterior cortex of the middle to proximal portion of the tibial shaft[249] (Fig 157). Radiographically, it is evidenced initially by periosteal reaction and then by an oblique fracture line that interrupts the cortex.

During the healing phase the periosteal new bone becomes well defined (Fig 158) and eventually becomes incorporated into the cortex, causing a slightly expanded shaft (Fig 159).

Proper diagnosis of this rather typical stress fracture is important since it may simulate the changes of bone neoplasm, osteomyelitis, or osteoid osteoma. Figures 160 and 161 show the extensive and expensive workup of similar cases, which might have been obviated by proper interpretation of the plain films.

Transverse Midshaft Fractures

Transverse fractures of the shaft tend to occur more frequently in the proximal portion and are seen commonly in military recruits[7] but may also occur in runners[8, 162, 250] and in children (see below). These fractures are more serious in that they have the potential for complete fracture, displacement, and even nonunion.[251] Consequently, early diagnosis and relief from activity are important. As in other stress injuries, the earliest confirmation is obtainable by bone scanning (Fig 162). The evolution of transverse fractures is identical to that in other tubular bones (Fig 163).

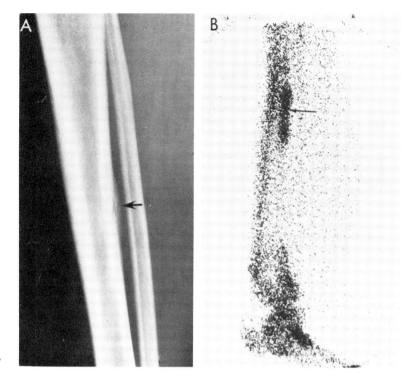

FIG 157.
A, typical runner's tibial stress fracture (*arrow*). **B,** nuclear scan confirms the fracture.

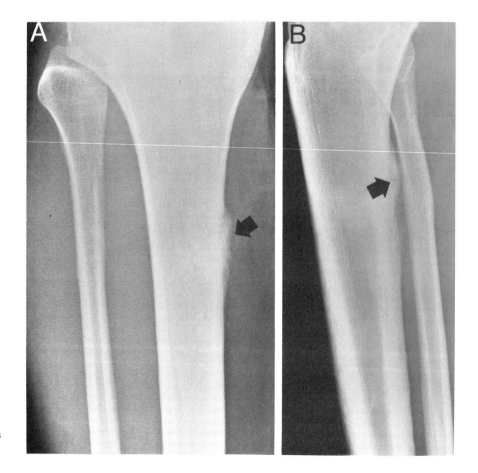

FIG 158.
A and **B,** healing runner's tibial stress fracture (*arrows*).

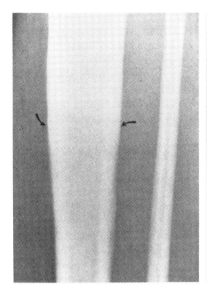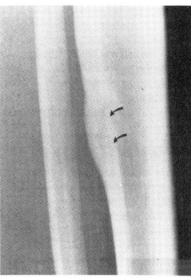

FIG 159.
A and **B,** healed runner's stress fracture of the tibia
(*arrows*).

Anterior Cortical Fractures

Radiolucent horizontal stress fractures of the anterior cortex of the tibia are typically seen in basketball players, professional dancers, and hurdlers, activities that require leaping in performance of athletic events.[252–254] These fractures have also been seen in a football kicker[255] and in gymnasts. The fracture is located in the middle and lower portions of the anterior cortex of the tibia (Fig 164). The location of these stress injuries is related to the muscular anatomy of the tibia. The powerful gastrocnemius and soleus supply plantar flexion of the foot, often forcefully, on leaping. During this action, the tibia acts as a bow and the soleus-gastrocnemius complex as a bowstring. Stretching the bowstring produces a corresponding bend or curve in the shaft of the bow. This stress is transmitted to the anterior cortex of the tibia.[252] With the passage of time, there is pronounced thickening of the anterior cortex of the tibia, which may be seen in both frontal and lateral projections (Fig 165). If untreated, some of these fractures may progress to complete fracture and nonunion.[253] Occasionally, these fractures may be seen bilaterally (Fig 166).

The lesion is readily diagnosable by plain radiography if one recognizes the entity. Figure 167 illustrates a case of a 15-year-old boy whose anterior cortical stress fracture was diagnosed as an osteoid osteoma and excised, leading to fracture through the operative site.

The anterior cortical stress fracture is often diagnosed clinically as "shin splints." This term is a nonspecific one used to describe a painful condition of the shin that interferes with training and performance in athletics. In many of these cases the symptoms are due to unrecognized stress reaction or stress fracture.[256] The American Medical Association's Standard Nomenclature of Athletic Injuries defines shin splints as "pain and discomfort in the legs from repetitive running on hard surfaces or forceful excessive use of the foot flexors" and states that "the diagnosis should

be limited to musculotendinous inflammation excluding fatigue fractures and ischemic disorders."[257, 258] Triple-phase bone scanning is recommended in all cases of shin pain to rule out stress reactions or fractures. Practically speaking, the term *shin splints* should be used to describe the patient's symptoms, and a search should be made for a more specific diagnosis.[259]

Fracture of the Distal Tibia

Fractures of the distal tibia are unusual in young people and may be seen in association with a stress fracture of the distal fibula.[8, 247] This location is much more commonly the site of an osteoporotic insufficiency fracture. These patients complain of ankle pain and swelling. Radiographically, these injuries are often recognized as compression stress fractures in that they are seen as bands of sclerosis extending across the shaft (Fig 168). A good history of overuse is difficult to obtain in many of these patients because they tend to be older and less involved in intense athletic activity.

Stress fractures of the distal tibia and calcaneus subsequent to ankle fractures of the tibia and fibula have been reported.[260] These patients were all between 5 and 38 years of age, had disuse osteopenia, and had recently begun weight bearing.

Longitudinal Stress Fracture

Longitudinal stress fractures are seen in very young children (the toddler's fracture) and in postmenopausal women, representing an insufficiency fracture. These injuries will be discussed in their appropriate sections below.

Stress Reaction

There is a continuum of stress injuries from stress reactions to stress fractures. Included in stress reactions

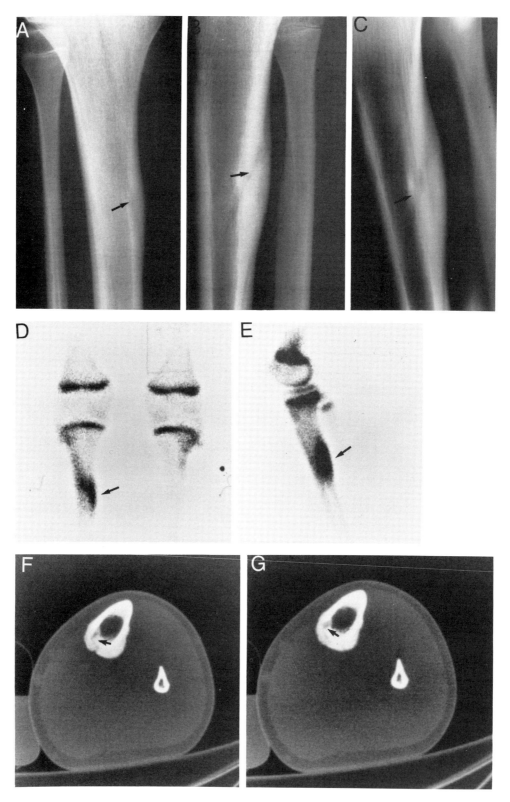

FIG 160.
A 19-year-old runner with a rather typical runner's fracture (*arrows*) incorrectly diagnosed as being an osteoid osteoma. **A** and **B,** plain films. **C,** tomogram. **D** and **E,** nuclear scan. **F** and **G,** computed tomographic sections showing the fracture of the posterior cortex and cortical thickening (*arrows*).

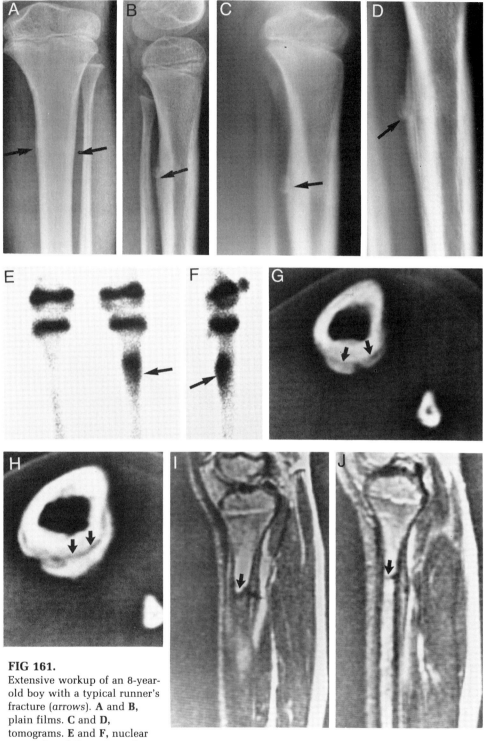

FIG 161.
Extensive workup of an 8-year-old boy with a typical runner's fracture (*arrows*). **A** and **B**, plain films. **C** and **D**, tomograms. **E** and **F**, nuclear scans. **G** and **H**, computed tomographic sections. **I** and **J**, magnetic resonance scans (1.5 T; TR$_2$; TE 80 msec).

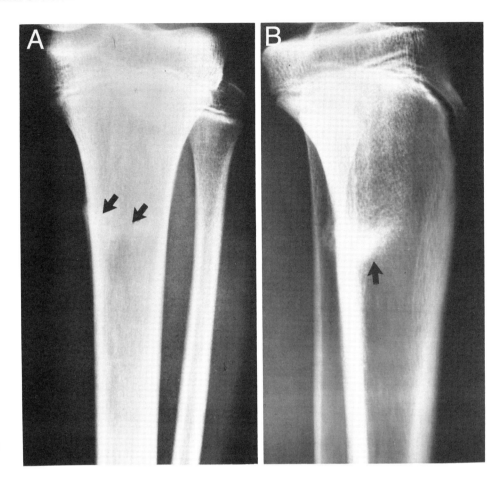

FIG 163.
A and **B,** transverse tibial shaft fracture in a 15-year-old athlete (*arrows*).

are those symptomatic patients who show areas of increased uptake on bone scans but who have normal roentgenograms and those patients who show only reactive periostitis on roentgenograms but have no discernible evidence of fracture (Figs 169 through 172). These kinds of reactions are common in athletes as well as in military recruits and are particularly important in patients with medial tibial pain, in whom periostitis may be the only radiographic evidence of stress.[261–263] Although nuclear scanning is useful in localizing the source of the patient's symptoms, it should be appreciated that initial bone scans may be normal early in the course of the disease but will become abnormal within weeks of the onset of symptoms. Repeated scans are therefore indicated in patients who continue to have symptoms after an initial normal examination result.[19, 264, 265]

Chronic stress reactions may also be documented by cortical thickening, as illustrated in Figure 173. This patient is a 29-year-old foundry worker who complained of left leg pain and whose radiographs showed thickening of the anterior cortex of the tibia. Questioning the patient indicated that his work entailed prolonged weight bearing on the left leg, which resulted in the evident reactive changes in the tibial shaft.

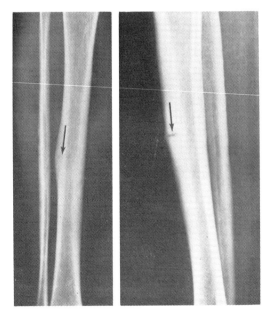

FIG 164.
A and **B,** anterior tibial cortical stress fracture in an 18-year-old gymnast (*arrows*).

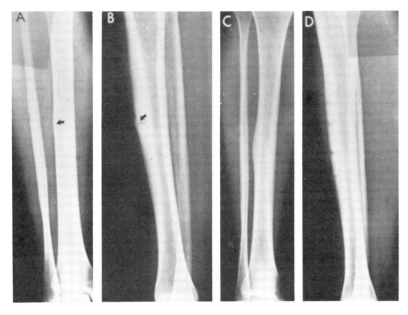

FIG 165.
A 16-year-old gymnast, runner, and cheerleader complained of pain in her right shin. Initial examination **(A** and **B)** showed cortical thickening in the anterior aspect of the midtibia and a horizontal cleft in **B** (*arrows*). Films made 3 months after cessation of athletic activity showed further increase in cortical thickening **(C)** and the presence of at least four anterior cortical stress fractures **(D)**.

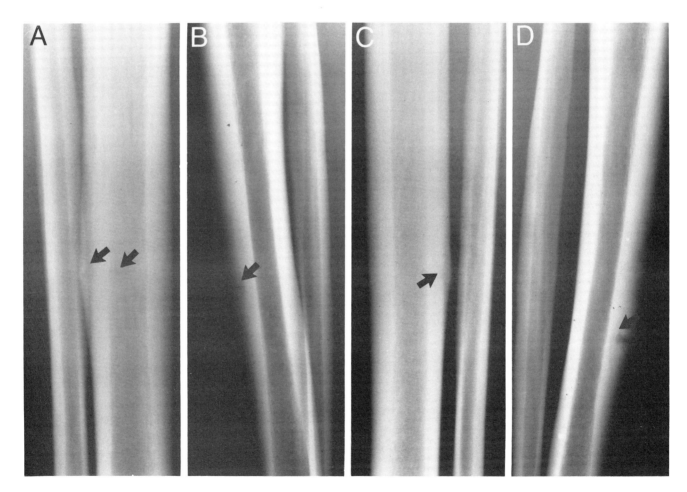

FIG 166.
Bilateral anterior tibial cortical stress fractures in a 27-year-old basketball player (*arrows*). **A** and **B,** right leg. **C** and **D,** left leg.

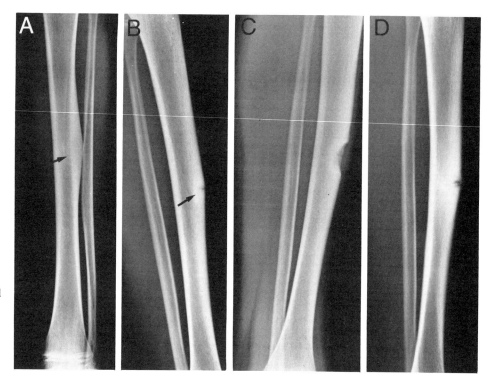

FIG 167.
A and **B,** typical anterior cortical tibial stress fracture (*arrows*) misdiagnosed as an osteoid osteoma. **C,** after excision. **D,** fracture through area of postsurgical weakness.

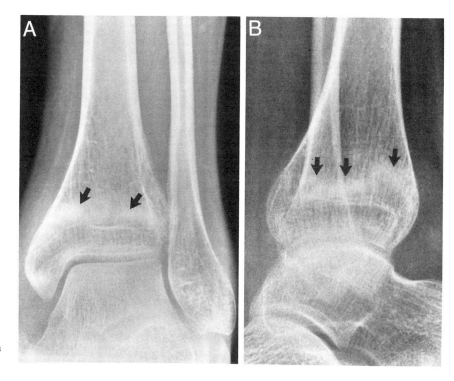

FIG 168.
A and **B,** stress fracture of the distal tibia in a 31-year-old man (*arrows*).

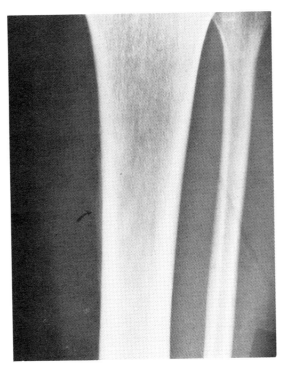

FIG 169.
Stress reaction of the tibia in a 19-year-old man (*arrow*).

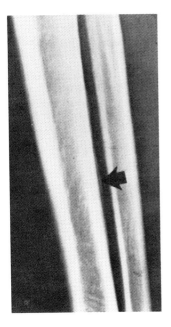

FIG 171.
Stress reaction of the posterior aspect of the tibia in a 20-year-old runner (*arrow*).

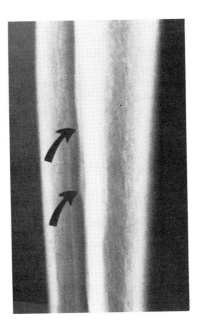

FIG 170.
Stress reaction in a 21-year-old athlete (*arrows*).

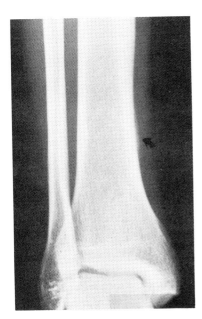

FIG 172.
Stress reaction of the tibia in a 16-year-old female runner (*arrow*).

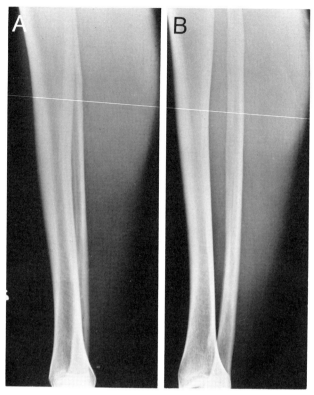

FIG 173.
A, anterior tibial cortical thickening in a 29-year-old man secondary to prolonged weight bearing on his left leg. **B,** comparison view of the right leg.

Stress Injuries of the Tibia in Children

Fractures

The most common stress fracture in young children is a transverse fracture of the proximal tibial shaft [8, 266–268] (Fig 174). When first seen, there is usually evidence of periosteal new bone, followed by radiolucency and/or sclerosis across the shaft of the bone. The difficulty in diagnosis often rests in the physician's failure to recognize a stress fracture in a young child. It is important that such recognition be made early to obviate biopsy, a procedure which can lead to an incorrect diagnosis of osteosarcoma. This concept is particularly important because the radiologic appearance of these stress fractures may mimic that of malignant neoplasia[269] (Fig 175). The clinical finding of localized pain, often subsiding with rest, is helpful evidence of stress fracture as opposed to a neoplasm.

The longitudinal tibial fracture of young children under the age of 3 years, the toddler's fracture, is considered by some investigators to be a stress fracture since there is often no appropriate history of trauma.[270, 271] The fracture is found in a child when weight bearing is just beginning. The child suddenly ceases to bear weight and becomes irritable. There is localized warmth and tenderness but no swelling. Radiographic examination shows soft-tissue edema initially and later an undisplaced fracture line that runs in a spiral or oblique manner inferiorly and medially (Figs 176 and 177).

Physiologic Bowing and Knock-Knee

Physiologic bowing and knock-knee represent normal

FIG 174.
A, frontal projection of a stress fracture of the tibia in a 6-year-old child (*arrow*). Fracture was initially thought to be an osteosarcoma. **B,** lateral projection. Note periosteal reaction.

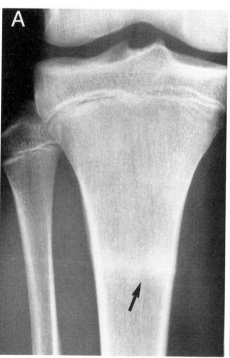
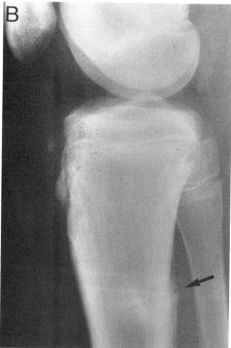

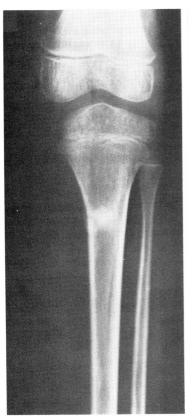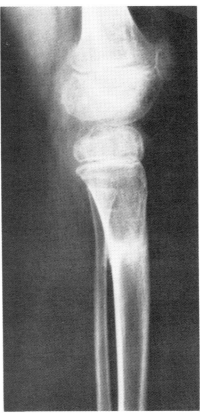

FIG 175.
A and **B,** stress fracture of the tibia in a 6-year-old
child. The injury was initially thought to be an
osteosarcoma.

phases of growth that are probably related to the stresses
of weight bearing and reflect the pliability of young
bone.[272, 273] Physiologically, bowing is seen in young children when weight bearing commences and lasts until the
age of 2 years. Between 2 and 12 years of age there is a
valgus phase. These normal physiologic events usually correct spontaneously (Figs 178 and 179). The associated
changes in the varus phase consist of thickening of the
medial cortex of the tibiae, beaking of the medial aspect
of the proximal tibial epiphysis and metaphysis, and similar changes in the distal femoral epiphysis and metaphysis. It has been noted that bowleg and knock-knee
deformities are more common in Jamaican children, possibly owing to earlier weight bearing.[274, 275]

Blount's disease (tibia vara) is considered to be an
extension of physiologic bowing that fails to straighten as
the child grows heavier.[274, 276, 277] The young child walks
with hips abducted and extremely rotated and knees flexed
on the tibiae, which have increased internal tibial torsion.
This places the weight on the posterior and medial portions
of the tibial plateau and results in a continuous deformity
and compressive force against this part of the epiphysis.
As upper tibial deformity occurs, there is increased shear
against the growth plate. Radiographically, the deformity
may be unilateral or at least asymmetric. The tibia is in
varus and the medial tibial metaphysis is depressed; the
bone is sclerotic and fragmented (Fig 180). Widening of

the lateral growth plate has been seen in some patients
with tibia vara.[278]

Avulsive Lesions

The Intercondylar Eminence.—Avulsion of the intercondylar eminence is predominantly a childhood injury,
most commonly the result of a fall from a bicycle. In the
adult, this fracture is the result of greater violence, and
there is usually extensive damage to supporting ligaments
and gliding surfaces of the knee joint. In most cases in
children, the fall on the bent knee is accompanied by simultaneous violent internal twisting of the tibia on the
femur.[279] A large effusion is usually present, and the fracture is quite evident. The fragment may remain in place
(Figs 181 and 182) or be separated from its point of origin
(Fig 183). This fracture occurs so frequently in bicycle ac-
cidents that it has been stated that a child between the ages
of 8 and 13 who has a painful swollen knee after falling
from a bicycle must be assumed to have this fracture until
proved otherwise. An osteochondral fracture of the tibial
plateau in a ballerina has been described resulting from
hyperextension during a twisting movement.[280]

The Tibial Tubercle.—Avulsion of the tibial tubercle
represents a Salter-Harris type III epiphyseal injury and
consists of a fracture through the physis underlying the
tibial tuberosity, with or without upward extension into

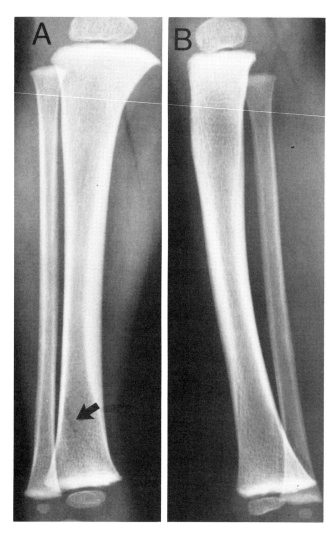

FIG 176.
A and **B,** toddler's fracture of the tibia in an 18-month-old child definitely seen only in frontal projection (*arrow*).

the contiguous proximal tibial epiphysis. It is seen primarily in boys between the ages of 14 and 16 years.[281] It is important to distinguish this injury from Osgood-Schlatter disease, in which the primary injury appears to be a partial avulsion of only the anterior portion of the ossification center of the tuberosity or of the epiphysis with no involvement of the physis.[282] The true avulsion fracture may involve only the tubercle portion of the epiphysis or may extend into the epiphysis of the tibial plateau with or without separation of the fragment or fragments (Fig 184). Premature closure of the epiphyseal plate is rarely seen as a complication of this injury. Occasionally, this fracture may be associated with ligamentous and meniscal tears.[283]

Osgood-Schlatter Disease.—Osgood-Schlatter disease is a disorder involving the growing tibial tuberosity in adolescents. It is believed to be the product of tearing of the

deep fibers of the patellar tendon at its insertion into the tibial tuberosity. The high tensions in this area are exaggerated by the additional stresses placed on it by active children. The average age of children with this complaint is 10 to 12 years. There is localized pain, swelling, and tenderness at the tibial tuberosity. There may be an increased incidence in children with valgus knees and low internally rotated patellas.[271]

The normal tibial tubercle may develop from multiple foci, and these nodules are not a sign of abnormality per se. The osseous changes of this disease consist of irregularity of the apophysis and of the contour of its component parts.[284] However, the soft-tissue signs have more diagnostic significance and include soft-tissue swelling over the tubercle, loss of definition of the patellar tendon, and edema of the infrapatellar fat pad[285] (Fig 185).

Later changes are variable. The fragments may unite, and the appearance of the tuberosity will be normal. Often, the dislocated fragments remain ununited and are seen anterior and superior to the tuberosity. The tuberosity itself may remain quite prominent with a bony mass extending into the patellar tendon (Fig 186).

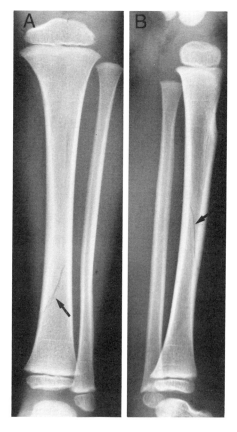

FIG 177.
Toddler's fracture of left tibia in a 3-year-old child (*arrows*). **A,** frontal projection. **B,** lateral projection.

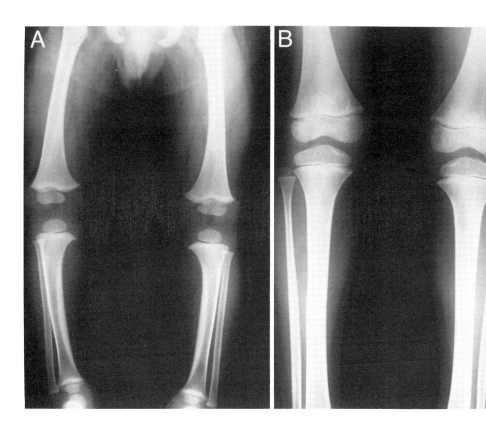

FIG 178.
A, physiologic bowing of the knees in a 2-year-old boy. **B,** spontaneous correction by age 4.

Patellar tendon avulsion has been reported as a complication of Osgood-Schlatter's disease brought about by activities involving rapid deceleration or vigorous acceleration.[286]

Ruptures of the Tibiofibular Interosseous Membrane.—Stress-induced rupture of the tibiofibular interosseous membrane will produce irregular calcification in the soft tissues between the fibula and tibia. These calcifications may not be seen for months or years following the avulsive episode (Fig 187). This elusive lesion may be seen in the absence of fracture of either the tibia or fibula (Fig 188).

Tug Lesions.—The soleal line on the posterior aspect of the proximal tibial shaft represents a point of one of the origins of the soleus muscle. In some individuals it may be exaggerated by muscle action and simulate an area of periostitis[287] (Fig 189).

The soleus muscle also arises from the posterior medial aspect of the neck of the fibula and may produce a tug lesion here as well, a lesion that may be mistaken for an osteochondroma (Fig 190). Both the soleal tug lesion and the fibular tug lesion may be seen in the same individual[146] (Fig 191).

THE FIBULA

Stress Fractures

The stress fracture of the fibula is usually located in the lower part of the fibula 3 to 8 cm above the top of the malleolus. However, in about 25% of cases, the fracture lies higher in the shaft.[8] These fractures are seen in runners,[8, 288, 289] military recruits doing jumping exercises,[290] ice skaters,[291] aerobic dancers,[292] volleyball, soccer, and squash players, and gymnasts.[288]

Stress fractures higher in the shaft are the result of jumping exercises[290] and running.[289] Low fractures tend to be transverse and the higher fractures oblique. It has been suggested that these fractures are caused by a combination of compression and torsion forces against the lateral malleolus or by the rhythmic contractions of the plantar and toe flexors.[288]

Pain and localized tenderness are characteristic, and in the lower fractures there is pain and stiffness of the ankle joint. As with other stress injuries, the pain is relieved by rest. In the early evolution, roentgenograms will demonstrate only periostitis (Fig 192). Later, the actual fracture may be seen by virtue of its lucency or condensation (Fig 193).

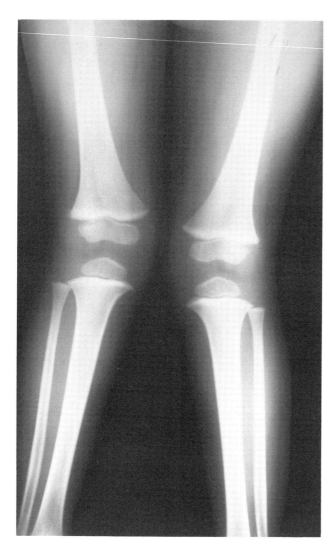

FIG 179.
Physiologic genu valgus in a 3-year-old boy. Note lack of any
architectural derangement.

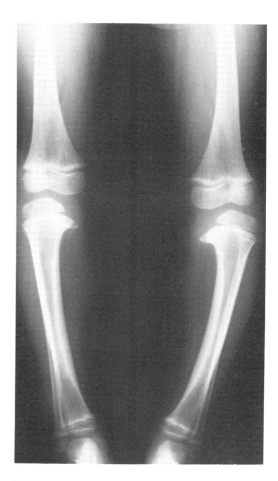

FIG 180.
Blount's disease.

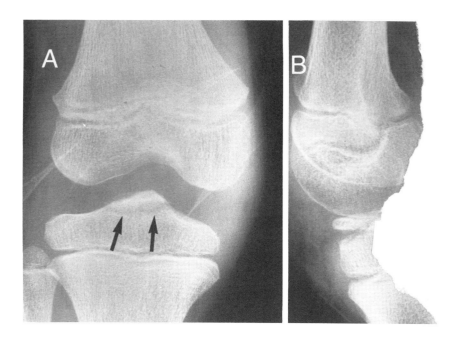

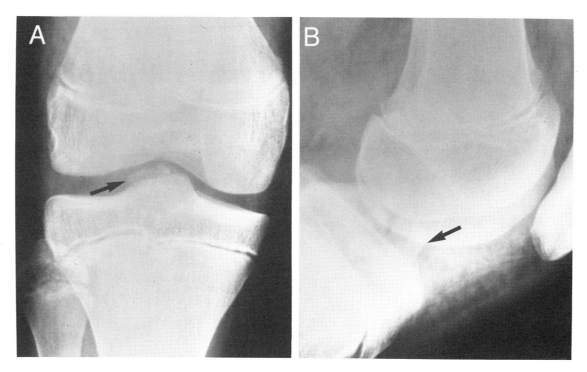

FIG 182.
Avulsion fracture of the tibial intercondylar eminence (*arrows*) in a 15-year-old boy who fell from his bicycle. **A,** frontal projection. **B,** lateral projection.

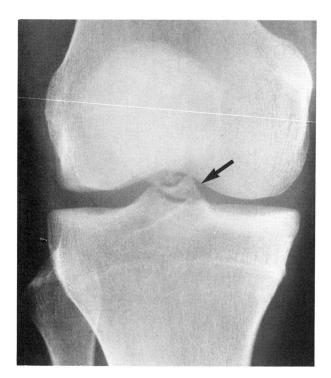

FIG 183.
Avulsion fracture of the intercondylar eminence (*arrow*) in a 17-year-old boy following a bicycle accident, with separation of the fragment.

Stress fractures of the distal fibula may be seen in association with stress fractures of the distal tibia at the same level and are usually of the condensing variety. Toddler's fractures of the fibula have been described and are manifested by a vague fracture line and subsequent periosteal proliferation in the distal shaft[8, 271] (Fig 194).

Epiphysiolysis

Stress-induced epiphyseal separations are more common in the upper extremity but are seen in the lower extremity as well. Figure 195 shows bilateral distal fibular epiphyseal separations in a 14-year-old long-distance bicyclist.

Avulsions

Avulsion fractures occurring at the tip of either malleolus are the result of pull mediated by the medial or lateral collateral ligaments and should be properly regarded as avulsion fractures rather than chip fractures, since there is no direct trauma (Fig 196). One should not confuse these avulsed fragments with ossicles, as ossicles are completely corticated and avulsed fragments are only partially corticated.

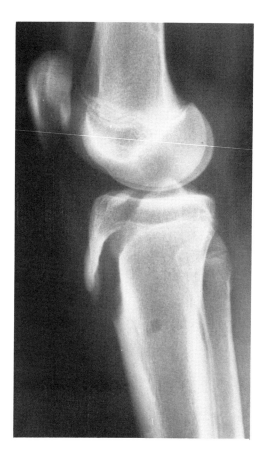

FIG 184.
Avulsion fracture of the tibial tubercle.

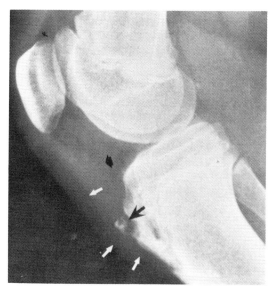

FIG 185.
Osgood-Schlatter disease of the tibial tubercle. Note loss of definition of the patellar tendon (*white arrows*) and edema of the infrapatellar fat pads (*arrowhead*), and fragmentation of the tubercle (*large arrow*).

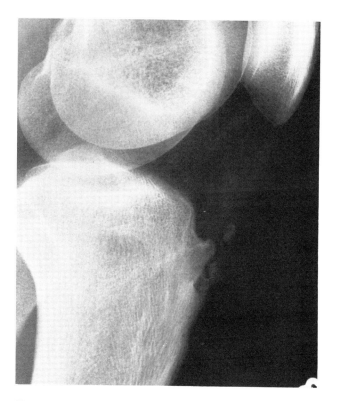

FIG 186.
Old healed Osgood-Schlatter disease with residual irregularity
and ossicles at the tibial tubercle.

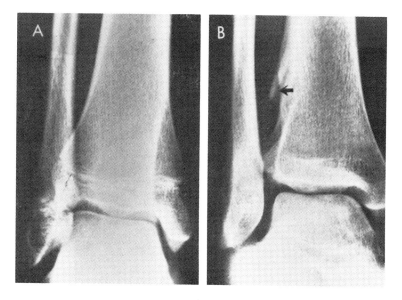

FIG 187.
A, baseline film made at age 17. **B,** film made at age 19 shows
stress-induced avulsive reaction at the interosseous membrane
insertion (*arrow*).

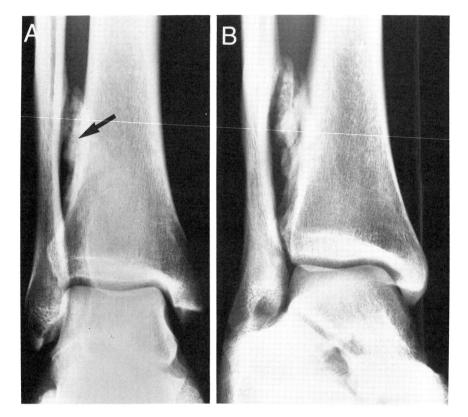

FIG 188.
Old healed avulsion of the interosseous membrane in a 25-year-old athlete (*arrow*). **A,** frontal projection. **B,** oblique projection.

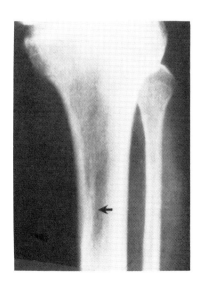

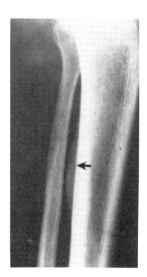

FIG 189.
A and **B,** the soleal line (*arrows*).

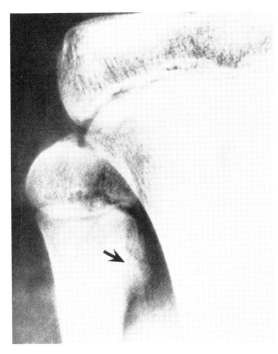

FIG 190.
Tug lesion at the origin of the soleus muscle in a 12-year-old boy (*arrow*).

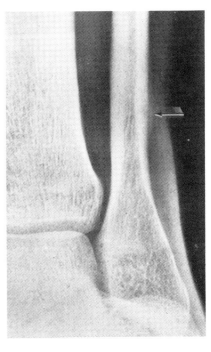

FIG 192.
Stress fracture of the fibula seen as a localized area of periosteal new bone formation in a 25-year-old runner (*arrow*).

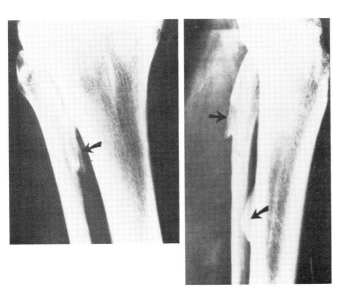

FIG 191.
A and **B,** soleal tug lesion and soleal line in the same individual (*arrows*).

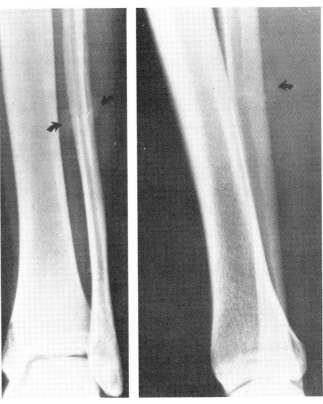

FIG 193.
A and **B,** stress fracture of the fibula in a 19-year-old basketball player (*arrows*).

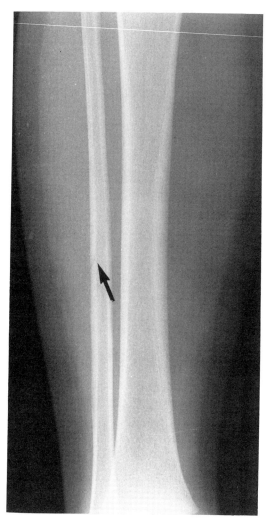

FIG 194.
Toddler's fracture of the fibula (*arrow*) in a 3-year-old boy with limp and no history of trauma. (From Ozonoff MB: *Pediatric Orthopedic Radiology*. Philadelphia, WB Saunders Co, 1979, p 463. Reproduced by permission.)

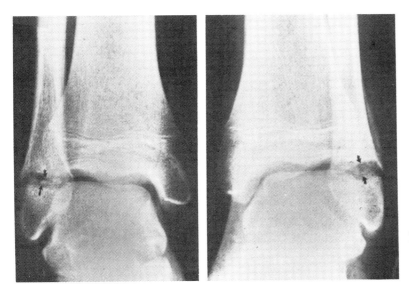

FIG 195.
A and **B,** stress-induced bilateral distal fibular epiphyseal separations in a 14-year-old long-distance bicyclist (*arrows*).

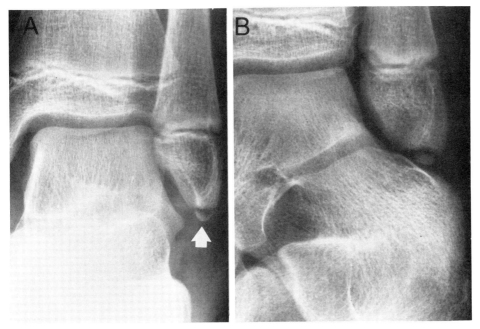

FIG 196.
Avulsion fracture of the tip of lateral malleolus resulting from pull of the lateral collateral ligament (*arrow*). **A,** frontal projection. **B,** oblique projection.

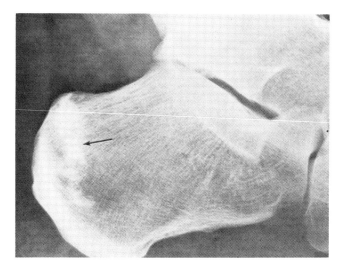

FIG 197.
Stress fracture of the calcaneus in a 22-year-old soldier seen as a band of condensation arranged in a linear configuration (*arrow*).

THE CALCANEUS

Stress Fractures

Stress fractures of the calcaneus are not uncommon in adults. It should be noted that the alterations seen in the calcaneal apophysis in adolescents, including irregularity and sclerosis of the apophysis which were formerly attributed to stress, are now generally accepted as growth phen-

oma rather than as indications of stress injury or osteochondrosis.

Stress fractures of the calcaneus are most commonly seen in soldiers or walkers.[7, 8, 293] Pain in the heel and swelling of the hind foot are usually evident. Confirmation by bone scanning will establish the diagnosis early in the course, but radiographic confirmation may take 1 to 5 weeks or even several months, particularly in older patients.

The classic radiographic finding is a patch or band of sclerosis in the middle or posterior portions of the calcaneus (Fig 197). Disruption of the superior cortex is not uncommon. Such fractures may be bilateral and symmetric or asymmetric[7, 294] (Fig 198). The sclerotic new bone usually extends from the junction of the body of the calcaneus to the tuberosity in its upper part, then downward and forward to the inferior surface. Displacement as a result of the fracture apparently does not occur.

Stress fractures of the calcaneus in children, similar to those seen in adults, are less common but are well documented and have been described as "another toddler's fracture." They have been seen between the ages of 19 and 41 months in normal children, who present with a limp.[295] These diagnoses are more readily made by bone scanning (Fig 199) but will eventually show the classic radiographic manifestations (Fig 200). Stress fractures of the calcaneus have also been described in children with spastic quadriparesis following periods of immobilization after orthopedic surgery[296, 297] (see Fig 200).

Stress fractures of the distal calcaneus have been described in association with other injuries and include stress fracture of the talar neck,[298] with stress fracture of the proximal shaft of the tibia.[299] These fractures have also occurred subsequent to immobilization for fracture of the tibia and fibula.[260]

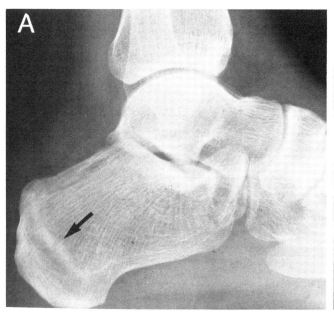
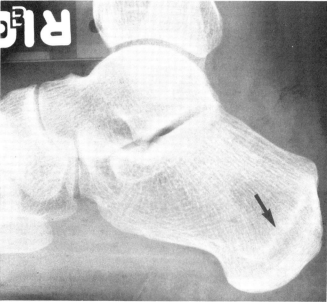

FIG 198.
Bilateral symmetric calcaneal stress fractures in a 25-year-old man (*arrows*). **A,** left heel. **B,** right heel.

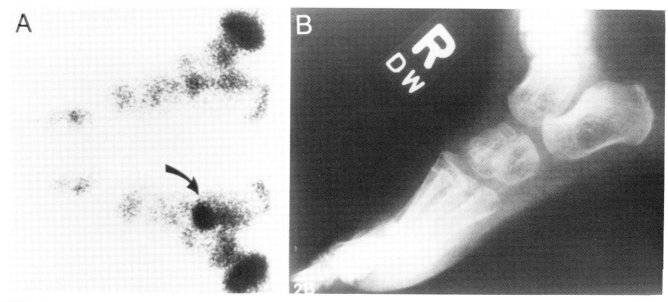

FIG 199.
A, lateral bone images of the feet. There is focal increased uptake in the distal aspect of the right calcaneus (*arrow*). **B,** lateral radiograph of the right foot is normal at this time. (From Starshak RJ, Simons GW, Sty JR: Occult fracture of the calcaneus: Another toddler's fracture. *Pediatr Radiol* 1984; 14:37. Reproduced by permission.)

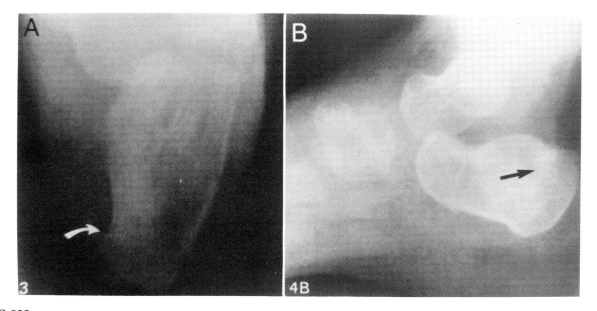

FIG 200.
A and **B,** examples of stress fractures of the calcaneus (*arrow*) in young children. (From Starshak RJ, Simons GW, Sty JR: Occult fracture of the calcaneus: Another toddler's fracture. *Pediatr Radiol* 1984; 14:37. Reproduced by permission.)

Stress Reactions

Stress reactions in the calcaneus are most unusual. Figure 201 illustrates a marked stress reaction in the calcaneus in a 20-year-old break dancer. This patient also had a stress reaction in his opposite femur, illustrated in Figure 115.

Tug and Avulsive Lesions

In healthy individuals, calcaneal spurs represent tug lesions. These are secondary to forces exerted by the plantar fascia and the Achilles tendon (Fig 202) and are of no clinical significance per se. They may become grossly enlarged in patients with diffuse idiopathic skeletal hyper-

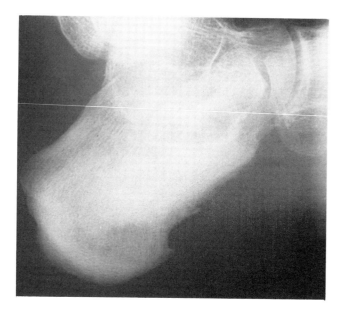

FIG 201.
Stress reaction of the calcaneus in a 20-year-old break dancer showing sclerosis of the inferior half of the calcaneus with irregular periosteal reaction on its plantar aspect. (From Ihmeidan IH, Tehranzadeh J, Oldham SA, et al: Case report 443. *Skeletal Radiol* 1987; 16:581. Reproduced by permission.)

ostosis. Figure 203 shows a common avulsive lesion of the calcaneus due to avulsion of the fibular collateral ligament secondary to inversion injury.

Haglund's syndrome is a common cause of posterior heel pain that results in a visible "pump-bump" at the posterosuperior border of the calcaneus. Haglund[300] associated the high incidence of this condition to the wearing of stiff, low-backed shoes while playing golf or hockey. The radiographic manifestations of this syndrome have been well described by Pavlov et al.,[301] and the reader is referred to this discussion for an in-depth treatment of calcaneal pitch lines that are used in making a radiologic diagnosis. The radiographic manifestations include a painful soft-tissue swelling at the insertion of the Achilles tendon, retrocalcaneal bursitis, thickening of the Achilles tendon, and an alteration in the configuration of the calcaneal pitch lines (Fig 204).

Rupture of the Calcaneal Tendon

Spontaneous rupture of the calcaneal tendon is a disease of the adult that may occur at any age but has its highest incidence in the 40- to 50-year-old, with a predominance in the male. The mechanism of injury in most cases is in the pattern of jumping, pushing off with the injured limb, or a significant fall. Most injuries occur dur-

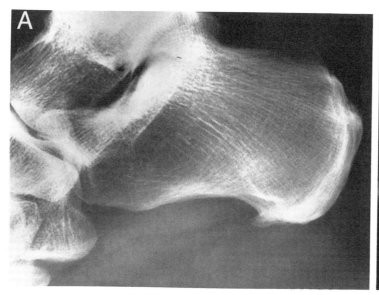
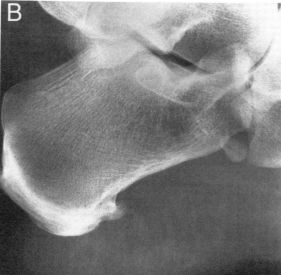

FIG 202.
A and **B**, two examples of typical calcaneal spurs.

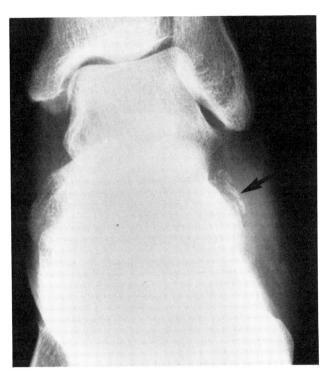

FIG 203.
Avulsion of the fibular collateral ligament (*arrow*).

ing some athletic pursuit or dancing. There is a decided relationship to treatment with steroids, and ruptures in this group of patients may be bilateral. There is also an association with rheumatoid arthritis, systemic lupus erythematosus, and local corticosteroid injection.[272]

Radiographic diagnosis is not particularly informative, although soft-tissue swelling and changes in tendon contour may be detected. Magnetic resonance imaging may dramatically demonstrate the tear[302] (Fig 205). Ultrasonography can also help determine the precise site and extent of the tear and document the healing response[303] (Fig 206).

THE TALUS

Stress Fracture

Talar neck stress fractures are very uncommon and are more likely to be diagnosed indirectly by bone scanning rather than by plain film. These fractures are usually seen in runners.[298, 304–306] The initial radiographic changes are seen as an area of increased linear density through the neck of the talus (Fig 207). The location in the neck of the bone appears to be quite characteristic.

Stress Reactions

The talar beak is considered a normal variation of the superior aspect of the talus (Fig 208). This beak represents the insertion of the joint capsule (Fig 209). In athletes,

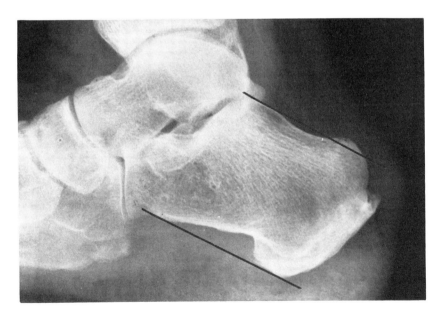

FIG 204.
Haglund's syndrome. There is loss of the retrocalcaneal recess, indicating retrocalcaneal bursitis, thickening of the Achilles tendon, loss of interface between the Achilles tendon and the pre-Achilles fat pad indicating Achilles tendonitis, and convexity of the posterior soft tissues at the level of the Achilles tendon insertion, indicating superficial tendoachilles bursitis (pump bump). (From Pavlov H, Heneghan MA, Hersh A, et al: The Haglund syndrome: Initial and differential diagnosis. *Radiology* 1982; 144:83. Reproduced by permission.)

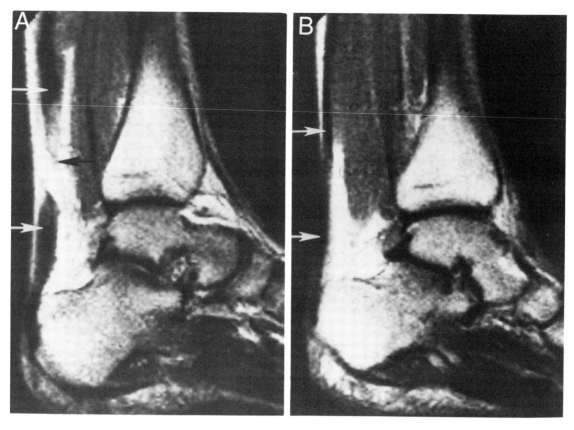

FIG 205.

A, sagittal magnetic resonance image (TE 1,600/40) shows the margins of the ruptured Achilles tendon (*white arrows*) as well as edema and hemorrhage between the tendon ends (*black arrow*). **B,** sagittal magnetic resonance image (TE 1,600/40) of the uninjured ankle demonstrates the normal course of the Achilles tendon (*arrows*). (From Reinig JW, Dorwart RH, Roden WC: MR imaging of a ruptured Achilles tendon. *J Comput Assist Tomogr* 1985; 9:1131. Reproduced by permission).

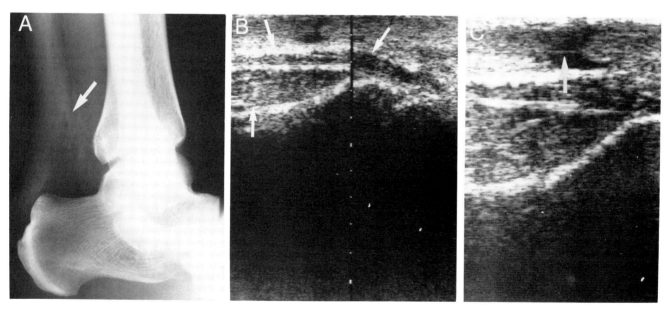

FIG 206.

A, lateral radiograph of ankle showing a small opaque defect (*arrow*) indenting Kager's lucent triangle. **B,** sagittal sonogram of posterior aspect of ankle. Achilles tendon rupture represented by lucent defect (*arrow 1*). Other visible structures are posterior margin of distal tibia (*arrow 2*), flexor hallucis longus muscle (*arrow 3*), and tendon of flexor hallucis longus (*arrow 4*). **C,** magnified sonogram of lucent tear traversed by thin echogenic line (*arrow*) representing strand of tissue. (From Leekam RN, Slasberg BB, Bogoch E, et al: Sonographic diagnosis of partial Achilles tendon rupture and healing. *J Ultrasound Med* 1986; 5:115. Reproduced by permission.)

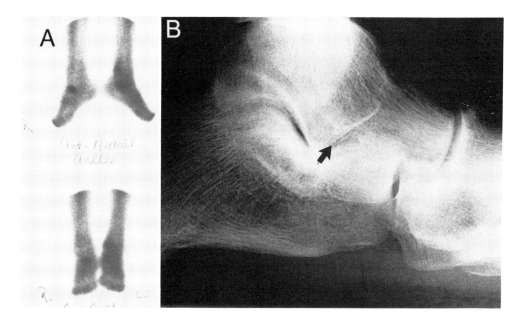

FIG 207.
A, talar stress fracture seen as an area of increased uptake of Tc-99m in the left talus. **B,** stress fracture of the neck of the talus seen as a linear density with a cortical break in the neck of the talus (*arrow*). (From Campbell G, Warnekros W: A tarsal stress fracture in a long-distance runner: A case report. *J Am Podiatry Assoc* 1983; 73:532. Reproduced by permission).

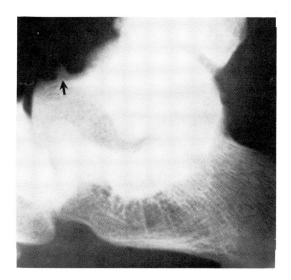

FIG 208.
The talar beak (*arrow*).

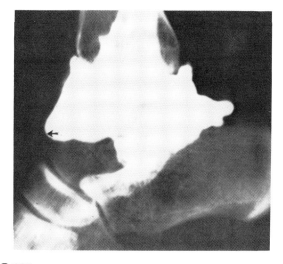

FIG 209.
Arthrogram of the ankle showing the anterior extent of the joint capsule at the site of the talar beak (*arrow*).

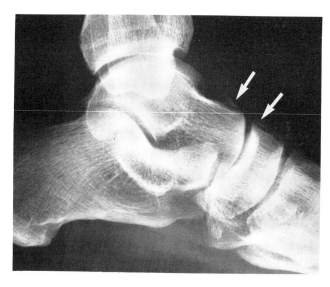

FIG 215.
Avulsion fractures of the superior aspects of the talus and navicular (*arrows*).

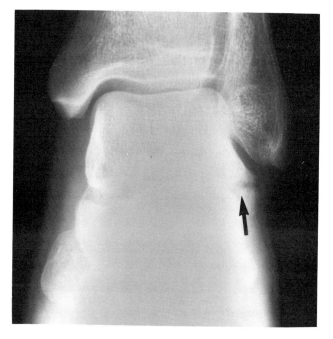

FIG 216.
Avulsion fracture of the lateral process of the talus (*arrow*).

THE NAVICULAR

Stress Fracture

Stress fractures of the tarsal navicular bone are often an unrecognized source of foot pain in athletes, particularly in runners, jumpers, and basketball and football play-

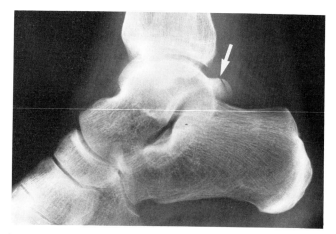

FIG 217.
Avulsion fracture of the posterior process of the talus (*arrow*).

ers.[317] These patients complain of ill-defined foot soreness or a cramping sensation that is increased during athletic activity. Tenderness is often present over the bone. Associated foot abnormalities may contribute to development of this lesion and include a short first metatarsal and metatarsus adductus as well as limited dorsiflexion of the ankle and/or limited subtalar motion. The presence of sclerosis of the proximal articular border of the tarsal navicular, narrowing of the talonavicular joint, talar beaking, accessory ossicles, and malalignment of the dorsal margins of the talonavicular and cuneonavicular joints in some patients may indicate the presence of some type of mechanical abnormality in the involved feet.[317–319]

Radiographic examination includes standing anteroposterior, lateral, and oblique projections of the foot. Coned anteroposterior views may be necessary to demonstrate the navicular. These fractures are often not seen in standard radiographs, and if these examination results are normal or equivocal, a radionuclide bone scan of both feet should be obtained. When the bone scan indicates a lesion of the tarsal navicular and plain films are normal, tomography should be carried out. Conventional anteroposterior tomograms are usually obtained with the sole of the foot flat on the table. In this position, the dorsal surface of the navicular is slanted with respect to the tomographic plane. In cases of suspected stress fracture, however, the forefoot should be elevated from the table with a sponge. In this position, the dorsal surface of the navicular is horizontal to the table and the tomographic plane. Fluoroscopic positioning is recommended. The foot is supinated until the medial facet of the navicular is seen en face, and the forefoot is elevated until the talonavicular joint is tangential to the beam.[319]

The fractures may be complete or incomplete and are usually in the sagittal plane in the central third of the navicular. Displacement may occur. The classic nuclear scanning and tomographic appearances are illustrated in Figures 218 and 219. Figure 220 shows an unusual stress fracture in the horizontal plane.

The Lower Extremity **121**

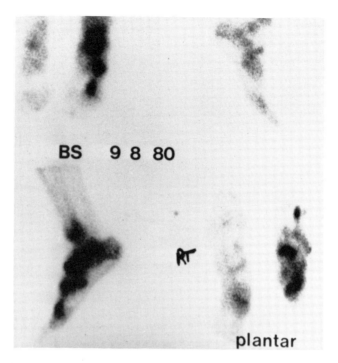

FIG 218.
Radionuclide bone scanning was performed starting in the
upper left quadrant and moving clockwise in the frontal, medial
(*right*), plantar, and medial (*left*) positions. There is augmented
isotope uptake in the left navicular and the fourth metatarsal. In
the frontal view, the tarsal area is overlapped by that of the
hindfoot, and it is difficult to localize the isotope uptake. On
the medial (*left*) view, the uptake is intensified in the region of
the navicular; however, it is poorly localized because of other
areas of augmented uptake. On the plantar view, the areas of
increased isotope uptake are best demonstrated and conform to
the configuration of the navicular and the fourth metatarsal,
respectively; stress fractures in both areas were documented.
(From Pavlov H, Torg JS, Freiberger RH: Tarsal navicular stress
fractures: Radiographic evaluation. *Radiology* 1983; 148:641.
Reproduced by permission.)

Osteochondrosis

The etiology of osteochondrosis of the tarsal navicular
is not clear, but there appears to be some agreement that
it is related to the stresses of weight bearing during growth
and can be considered a stress lesion in this sense. There
is good evidence that the disease is secondary to interfer-
ence with the vascular supply of the bone.[320]

The condition is more common in boys than in girls
and is seen in young children, with a peak incidence of 5
to 6 years. Pain and swelling are often present radiographically. The lesion is characterized by compression and increase in density leading to a disc-like configuration (Fig
221). The normal irregularity and apparent fragmentation
of the developing tarsal navicular should not be misconstrued as evidence of osteochondrosis.[146] Recovery occurs
with eventual restoration of the normal configuration of
the bone.

Avulsions

Avulsion fracture of the tarsal navicular is not uncommon and is most often seen at the superior aspect of the
bone, at times associated with avulsion fracture of the talus
(Fig 222). Care should be taken not to mistake these small
avulsions for the accessory ossicle, the os supranaviculare[146]
(Fig 223). Another avulsion fracture that may resemble an
accessory ossicle is the avulsion fracture of the medial end
of the navicular, which simulates the os tibiale externum[146]
(Fig 224). Lack of complete cortication of avulsed fragments together with localized signs and symptoms are useful clues in differential diagnosis.

FIG 219.
A, an anatomic anteroposterior
(AP) tomogram through the dorsal
aspect of the navicular
demonstrates a partial tarsal
navicular stress fracture. The
fracture is linear and sagittal and
interrupts the proximal articular
border. Note the sclerosis of the
proximal articular border of the
navicular. **B,** an anatomic AP
tomogram through the dorsal
aspect of the navicular
demonstrates a partial stress
fracture involving the entire dorsal
surface and interrupting both the
proximal and distal articular
borders. (From Pavlov H, Torg JS,
Freiberger RH: Tarsal navicular
stress fractures: Radiographic
evaluation.*Radiology* 1983;
148:641. Reproduced by
permission.)

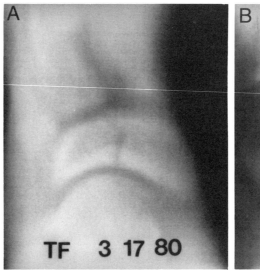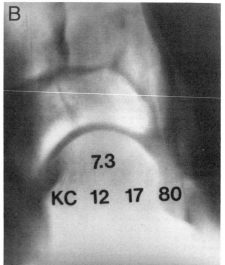

FIG 220.
A and **B,** unusual stress fracture of
the navicular in the horizontal
plane in a 25-year-old runner
(*arrow*). The fracture is not seen in
the anteroposterior plane. This
view also shows an os tibiale
externum (*arrow*).

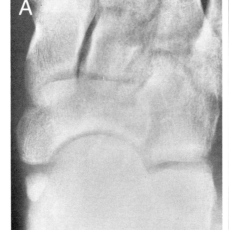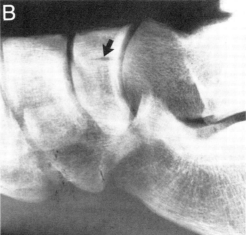

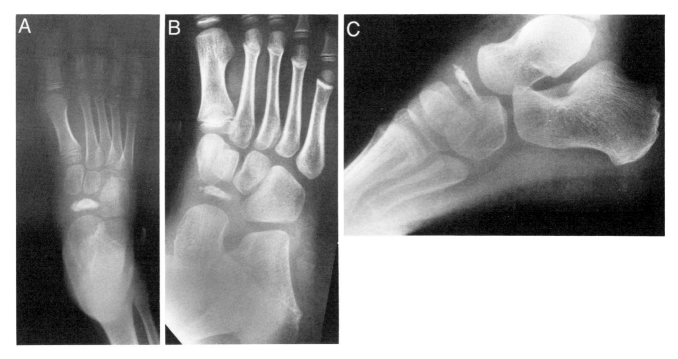

FIG 221.
Kohler's disease of the tarsal navicular in a 7-year-old boy. **A,** anteroposterior projection. **B,** oblique projection. **C,** lateral projection.

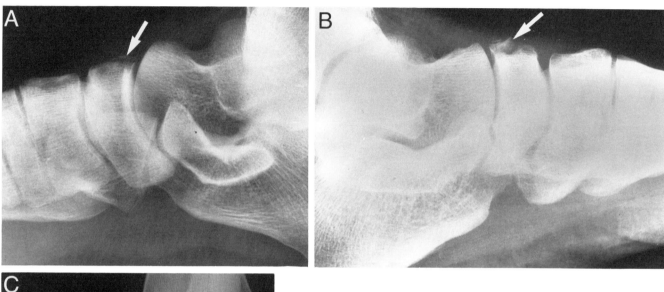

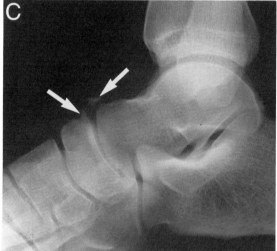

FIG 222.
A and **B,** avulsion fractures of the superior aspect of the
navicular (*arrow*). **C,** avulsion fracture of the superior aspect of
the navicular with an associated avulsion fracture of the talus
(*arrows*).

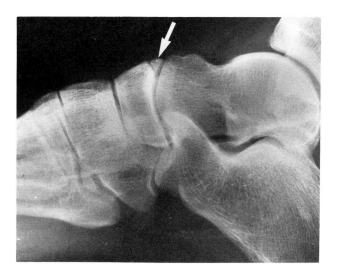

FIG 223.
Avulsion fracture of the tarsal navicular (*arrow*) simulating an os supranaviculare.

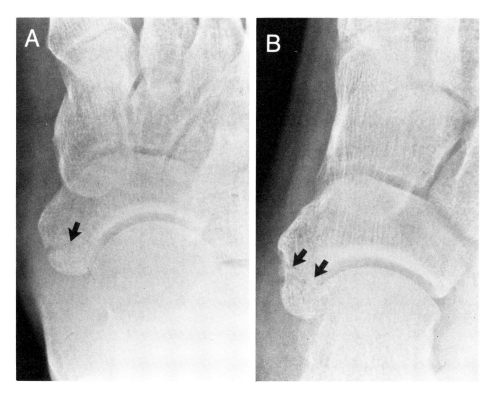

FIG 224.
Avulsion fracture of the medial end of the tarsal navicular simulating the os tibiale externum (*arrows:* compare with Fig 220). **A,** frontal projection. **B,** oblique projection.

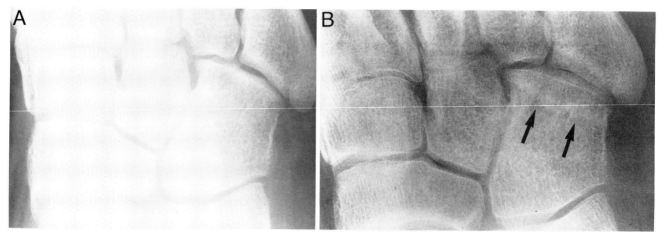

FIG 225.
Stress fracture of the cuboid. **A,** initial examination shows no abnormality. **B,** examination 1 month later shows a stress fracture of the distal portion of the bone (*arrows*). (Courtesy of Dr. Murray K. Dalinka).

THE CUBOID

Stress fractures of the cuboid are very rare; consequently, there are few reports in the literature.[162] A single case in an 18-month-old girl was reported and presumably was the result of early weight bearing.[321] Figure 225 illustrates a case of a stress fracture in a young adult.

THE CUNEIFORMS

Stress fractures of the cuneiforms are unusual and have been seen almost exclusively in soldiers.[322, 323] Stress fracture of the first cuneiform was described by Maseritz in 1936[324] and of the second cuneiform by Childress in 1943.[325] Stress fracture was described in the third cuneiform by Meurman and Elfving in 1980.[323] The findings are similar to those of stress fracture in other tarsal bones and are illustrated in Figures 226, 227, and 228.

THE METATARSALS

Stress Fractures

Stress fractures of the metatarsals (march fractures) are the most frequently encountered injury of this type. Although it is most commonly seen in the military,[326, 327] it is also seen in others who walk a great deal, such as postmen, athletes, and ballet dancers,[328] and following operations for bunion with hallux valgus or hallux rigidus.[329]

The predominant symptom is pain on prolonged weight bearing, followed in 3 to 10 days by swelling of the dorsum of the foot, occasionally accompanied by erythema.[7] The most common fractures are those that occur in the second or third metatarsal shafts.

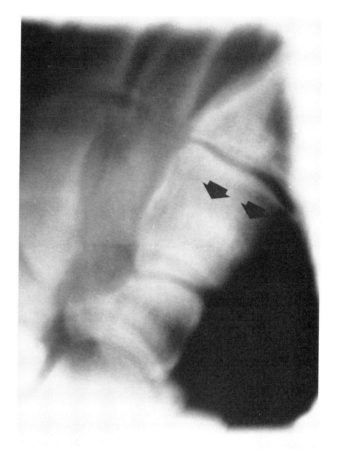

FIG 226.
Stress fracture of the first cuneiform (*arrows*) demonstrated by tomography. (From Meurman KO, Elfving S: Stress fracture of the cuneiform bones. *Br J Radiol* 1980; 53:157. Reproduced by permission.)

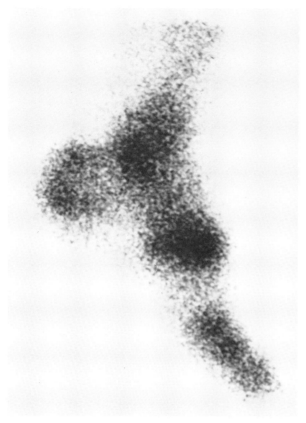

FIG 227.
Nuclear scans of a patient with a stress fracture of the third right cuneiform showing increased activity in the medial tarsometatarsal region. (From Meurman KO, Elfving S: Stress fracture of the cuneiform bones. *Br J Radiol* 1980; 53:157. Reproduced by permission.)

The earliest radiologic signs consist of a cleft of the bone or, more commonly, a poorly defined area of periosteal new bone, which is not usually circumferential (Fig 229). With the passage of time, the periostitis becomes more exuberant, and one may see a cleft through the shaft (Fig 230). Eventually, the periosteal new bone is incorporated with the cortex and becomes better defined (Fig 231). Displacement of the bone fragments is unusual and is more likely to be seen in older patients with osteoporosis.[8]

An unusual form of metatarsal stress fracture has been described in a 16-year-old long-distance runner.[330] This was a Salter-Harris type II epiphyseal injury at the base of the first metatarsal. All stress fractures of the first metatarsal are rare, but a more common variation is that through the base of the first metatarsal seen as a band of sclerosis across the shaft with periosteal reaction[7, 326] (Fig 232).

Stress fractures of the fourth metatarsal are rare and are usually seen in conjunction with multiple stress fractures involving the other metatarsals.

Stress fractures of the fifth metatarsal are of two types. The first occurs at the base with the fracture line opening on the lateral side and running transversely across the bone (Fig 233). The second type is much less common and occurs in the shaft in the same pattern as those in the other metatarsals, but in conjunction with other metatarsal stress fractures.[7, 8, 317, 331]

Stress Reactions

The classic stress reaction in the foot is the entity known as Morton's foot, in which the shaft of the second metatarsal becomes enlarged and its cortex thickened, usually in response to increased stress of weight bearing in-

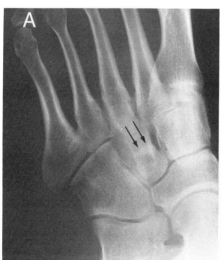

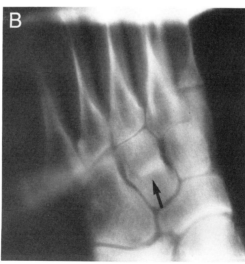

FIG 228.
Stress fracture of the third cuneiform (*arrows*). **A,** oblique projection. **B,** tomogram. (From Meurman KO, Elfving S: Stress fracture of the cuneiform bones. *Br J Radiol* 1980; 53:157. Reproduced by permission.)

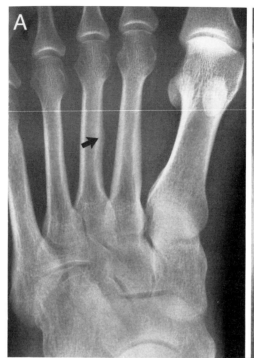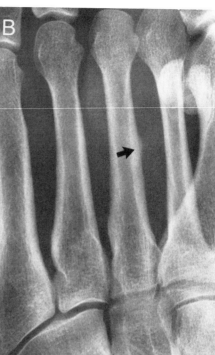

FIG 229.
Early stress fracture of the third metatarsal (*arrows*) seen as a small area of new bone formation at the medial aspect of the midshaft in a 31-year-old aerobic dancer. **A,** frontal projection. **B,** oblique projection.

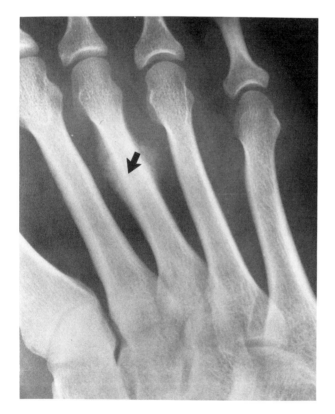

FIG 230.
Stress fracture of the third metatarsal seen as an area of periostitis with a subtle cleft (*arrow*) through the shaft in a 20-year-old soldier.

duced by a short first metatarsal (metatarsus atavicus). I have seen the same phenomenon in cases in which the third and fourth toes are relatively short (Fig 234). These configurations may be seen in asymptomatic individuals as well as in those with symptoms.[332] A similar phenomenon may be seen in the second metatarsal following osteotomy on the first metatarsal.[333] Cortical thickening of the metatarsals, particularly of the second and third metatarsals, has also been described in ballet dancers as a result of chronic stress.[334]

Osteochondrosis

Osteochondrosis of the heads of the metatarsals with or without osteochondrosis dissecans has been related to chronic stress of weight bearing and is presumed to be a type of aseptic necrosis (Freiberg's infraction). The disease usually involves the second or third metatarsal heads. It occurs most commonly in females, usually beginning in adolescence.[335] Figures 235 and 236 show typical examples of this disease. Figure 237 illustrates the osteochondrosis dissecans of the first metatarsal head. Figure 238 shows osteochondrosis dissecans of the second metatarsal head superimposed on Morton's foot with the classic enlargement of the second metatarsal associated with a short first metatarsal. These kinds of cases lend weight to the concept that the etiology of the disease is related to faulty weight bearing and stress.

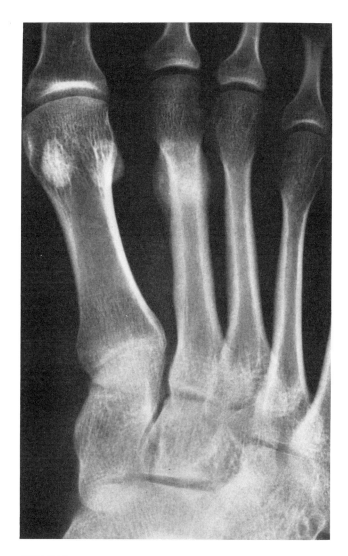

FIG 231.
Healed stress fracture of the second metatarsal.

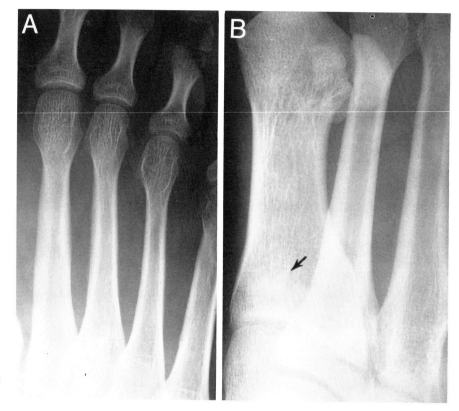

FIG 232.
Stress fracture through the base of the
first metatarsal with sclerosis of the
fracture line and periosteal reaction
(*arrow*). **A**, frontal projection. **B**, oblique
projection. (From Levy JM: Stress
fractures of the first metatarsal. *AJR* 1978;
130:679. Reproduced by permission.)

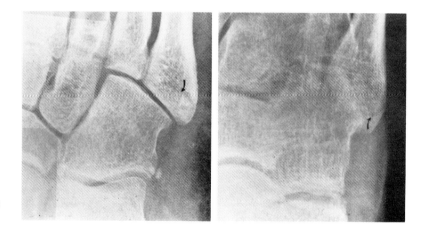

FIG 233.
A and **B**, stress fracture of the base of the fifth
metatarsal in a 21-year-old runner, without bone
reaction (*arrows*).

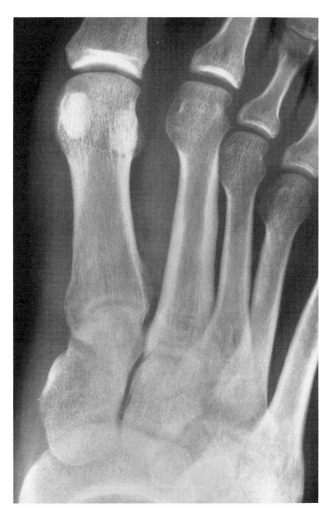

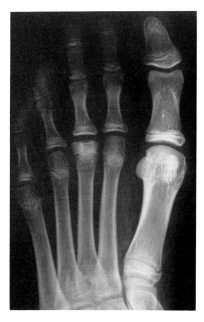

FIG 235.
Typical Freiberg's disease of the third metatarsal head in a 16-year-old girl.

FIG 234.
Morton's foot with compensatory enlargement of the shaft of the second metatarsal accompanied by thickening of the cortex. Note short third and fourth metatarsals.

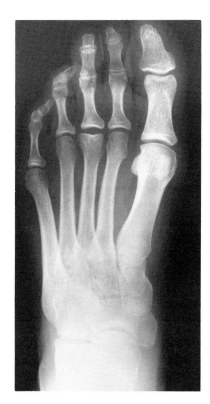

FIG 236.
Healed Freiberg's disease of the third metatarsal head in a 25-year-old woman.

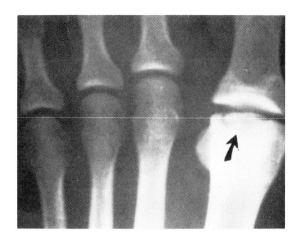

FIG 237.
Osteochondrosis dissecans of the head of the first metatarsal (*arrow*).

THE TARSAL SESAMOID BONES

Stress Fractures

Stress fractures of the tarsal sesamoid bones are usually seen at the head of the first metatarsal. Absence of previous injury, patient history of athletic effort, and the lack of history of acute onset of pain will usually exclude ordinary fracture. Diagnosis is usually made from the history and by the presence of localized pain over the sesamoids and can be confirmed by nuclear scanning[336, 337] and/or radiography.[162, 338–340] Radiographic diagnosis may be complicated by the fact that these sesamoids show wide variation in their development and may be multipartitioned.[341] In cases of true or stress fractures, the fragments usually do not fit together, and the margins along the radiolucency are sharp and not beveled. This is best seen in tangential projections of the sesamoids similar to the tangential or "sunrise" view of the patella.[162] Stress fractures

may occur in either the medial or lateral sesamoids. Figures 239 and 240 show stress fractures of the medial sesamoid in runners.

Bone scans may be helpful in cases in which there is difficulty in differentiating a partitioned sesamoid from a fracture (Fig 241).

Fracture of the Os Peroneum

The os peroneum is a sesamoid in the tendon of the peroneus longus muscle. Rarely, this ossicle will fracture as a result of an avulsive type of injury due to dorsiflexion of the foot[342] or as the result of stress injury.[343] The injury often resembles a sprain, but the radiographic findings are quite typical and show the division of the ossicle with sharp margins and lack of cortication along the line of division (Fig 242). Separation of the fragments indicates rupture of the tendon at the level of the sesamoid bone.[344] Care must be taken not to confuse this fracture with the congenitally partitioned os peroneum.[146]

Osteochondrosis

Osteochondrosis or aseptic necrosis of the sesamoids of the foot may occur without physical stress or trauma other than weight bearing and is commonly localized in the medial bone; the condition is more frequent in females. There is some suggestion that stress fracture and osteochondrosis may be identical.[340] The disease has been seen in marathon runners, baseball players, and ballet dancers.[162, 345]

Scintigraphic examination shows increased uptake in the involved areas of the great toes, and radiographic examination shows fragmentation, mottled density, and irregularity of the sesamoid (Figs 243 and 244). These changes are often better seen in tangential projection. Excellent symptomatic results are obtained by excision of the affected sesamoid.[346]

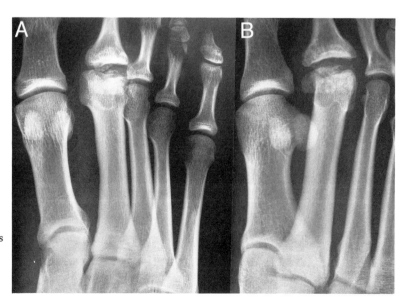

FIG 238.
Osteochondrosis dissecans superimposed on Morton's foot with stress changes evident in the second metatarsal shaft. **A,** frontal projection. **B,** oblique projection.

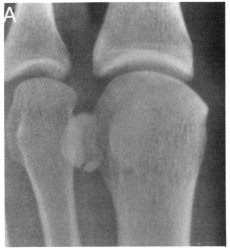

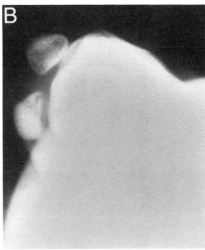

FIG 239.
Stress fracture of the medial sesamoid of the great toe in a 25-year-old runner. **A,** oblique projection. **B,** tangential projection.

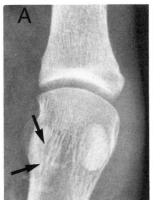

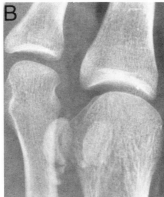

FIG 240.
Stress fracture of the medial sesamoid of the great toe in a 20-year-old runner (*arrows*). **A,** frontal projection. **B,** oblique projection.

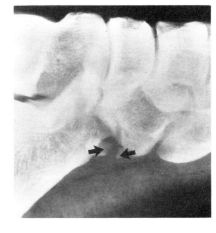

FIG 242.
Avulsion fracture of the os peroneum (*arrows*).

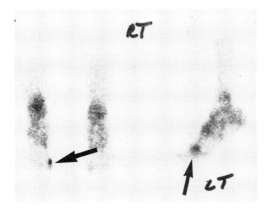

FIG 241.
Nuclear scan in a 30-year-old patient with stress injury of the sesamoid of the great toe showing localized increased uptake (*arrows*) in the area of the sesamoid of the left big toe. (From Chillag K, Grana WA: Medial sesamoid stress fracture. *Orthopedics* 1985; 8:819. Reproduced by permission.)

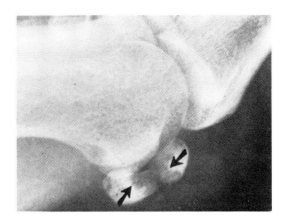

FIG 243.
Osteochondrosis of the sesamoids of the first toe (*arrows*).

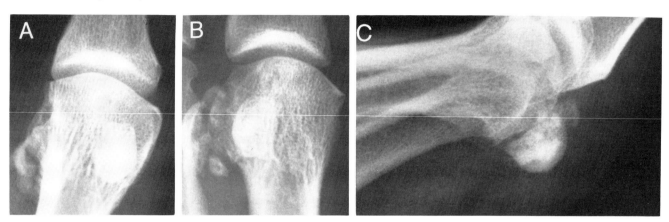

FIG 244.
Osteochondrosis of the medial sesamoid of the first toe. **A,** frontal projection. **B,** oblique projection. **C,** lateral projection.

PART II

Stress Injuries of the Abnormal Skeleton

Stress or fatigue fractures of abnormal bone are termed *insufficiency fractures*, a term that denotes a fracture in bone that possesses less than normal resistance to stretching and compression. The term *insufficiency fracture* is more meaningful than many of the terms used previously and is less confusing than other applied terms such as *pseudofracture, spontaneous fracture,* or *pathologic fracture.*[347]

Insufficiency fractures occur in bone with diminished volume (osteopenia). Many diseases may lead to osteopenia and eventually to insufficiency fracture (Table 9). Since the most frequent types of osteoporoses are postmenopausal and senile osteoporosis, insufficiency fractures are most frequently encountered in the elderly. The types of insufficiency fractures will be discussed below in accordance with their etiology and anatomic locations.

TABLE 9.

Diseases and Conditions Associated With Osteopenia*

Postmenopausal osteoporosis (women)
Senile osteoporosis (men)
Rheumatoid arthritis
Reconstructive hip and knee surgery
Steroid therapy
Radiotherapy
Secondary hyperparathyroidism
Osteomalacia
Paresis due to nerve injury
Amenorrhea

*From McNiesh LM: Unique musculoskeletal trauma. *Radiol Clin North Am* 1987; 25:1107. Reproduced by permission.

8

Osteoporosis

Osteoporotic bone is particularly susceptible to stress fracture. These fractures create a diagnostic problem because there is usually no history of excessive activity. Often, careful history taking will reveal a change in the usual pattern of activity with the addition of a seemingly minor alteration, either in the type of daily activity or its duration. These fractures are difficult to diagnose correctly, often being interpreted radiologically as neoplastic metastatic deposits, particularly since these fractures are seen predominantly in middle-aged postmenopausal or elderly women. Other susceptible individuals are those who are being treated with corticosteroids for other diseases and have developed osteoporosis as a result of the medication.

The pain of insufficiency fracture is variable. It may be continuous or absent entirely and does not bear the close association with exercise evident in the stress fracture of the young individual.

Insufficiency fractures present radiologically as a band of sclerosis in trabecular bone and as a lucent fracture line with or without periosteal reaction in cortical bone. As in all stress injuries, the scintigraphic findings will precede the radiographic findings.

THE SPINE

Compression fracture is probably the best-appreciated manifestation of injury to the osteoporotic spine. It is evident in postmenopausal women as well as in patients with senile osteoporosis. Similar fractures are also seen in patients receiving cortisone therapy[348] and are evidenced by radiolucency of the vertebral bodies, wedging of the vertebrae, and kyphosis. The vertebral end-plates are thin and often show a concavity producing the "cod fish vertebra" configuration.[349] The typical radiographic findings are shown in Figures 245 and 246. Occasionally, these fractures may resemble burst-type fractures and may have retropulsed fragments best seen by tomography or magnetic resonance imaging. In such cases, an abnormal posterior vertebral body line may be the only abnormality detected by conventional radiography.[350]

Osteopenia may be seen in childhood leukemia and also as a by-product of its treatment with steroids. The disease and/or its treatment may also lead to multiple vertebral compression fractures without trauma and result in a characteristic appearance of osteopenia and compression fracture of the vertebrae at multiple levels[351, 352] (Fig 247). Insufficiency fractures of the vertebrae resulting from steroid medications have an additional manifestation of bony condensation at the end-plates of the vertebrae, which is quite striking in contrast to the osteoporotic vertebral body[353] (Fig 248).

THE STERNUM

The kyphosis that results from osteoporotic compression fracture of the dorsal spine creates a deforming stress on the sternum that may, in turn, lead to an insufficiency stress fracture of the body of the sternum[354] (Fig 249).

THE UPPER EXTREMITY

Patients with anorexia nervosa may have associated hypothalamic amenorrhea, hypoestrogenemia and diminished bone mineral density, and stress fractures. Such a case has been reported by Baum et al.[355] of a 25-year-old woman who engaged in weight lifting and developed bilateral stress fractures of her ulnae.

Osteoporotic fractures of the spine, pelvis, and humerus have also been reported in patients with anorexia nervosa[356–358] as well as in others with prolonged amenorrhea.[328]

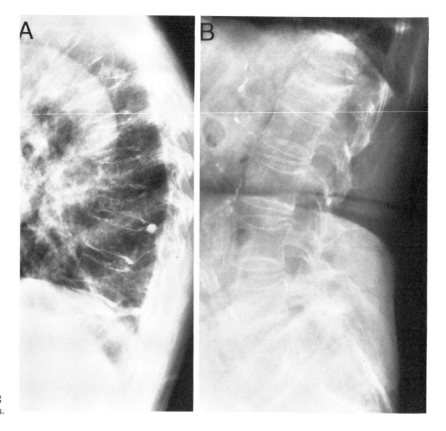

FIG 245.
A and **B,** senile osteoporosis of the dorsal and
lumbar spine in a 78-year-old woman showing
osteopenia and multiple compression fractures.

THE SACRUM

Insufficiency stress fractures occur in the sacrum after
radiotherapy or secondary to postmenopausal or steroid-
induced osteoporosis. Findings on plain films and conventional tomograms are often subtle and, as a result, are
often either overlooked or confused both clinically and
radiographically with metastatic disease. Radionuclide bone
scans show a characteristic distribution of increased uptake. Computed tomography is the definitive technique for
demonstrating the fracture.[32, 35, 359–363]

Because of the diffuse involvement of the skeleton by
osteoporosis, the possibility of multiple concurrent insufficiency fractures should be recognized. Especially common combinations include the sacrum and pubis,[359, 363] the
sacrum and supra-acetabular ilium, and the pubis and supra-acetabular ilium.[32]

Insufficiency fractures of the sacrum are often an unsuspected cause of low back, hip, and leg pain in elderly
women. Symptoms suggestive of a cauda equina syndrome
may be present, but there is minimal or no neurologic
deficit. There is usually marked sacral tenderness.[360]

One of the most striking features of these sacral fractures is their consistent location. The fractures course ver-
tically in the sacral alae, parallel to the sacroiliac joints,
just lateral to the margins of the lumbar spine.[32] There may

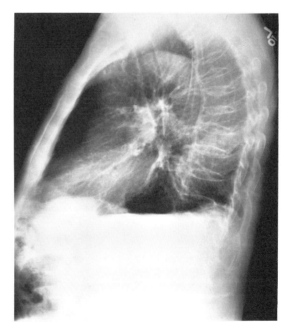

FIG 246.
Postmenopausal osteoporosis of the dorsal spine in a 62-year-
old woman with multiple compression fractures and bowing of
vertebral end-plates.

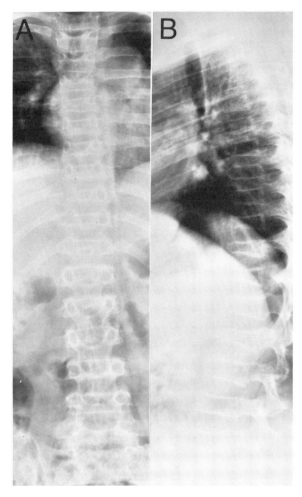

FIG 247.
Osteopenic compression fractures of the spine in a 6-year-old boy with leukemia. **A,** frontal projection. **B,** lateral projection.

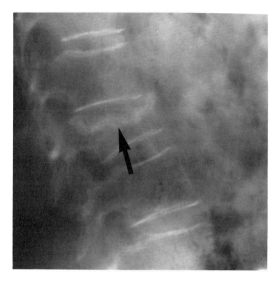

FIG 248.
Cortisone-induced osteoporosis of the lumbar spine (*arrow*) in a 30-year-old woman with insufficiency fracture of L-2 with condensation of bone at the fracture site.

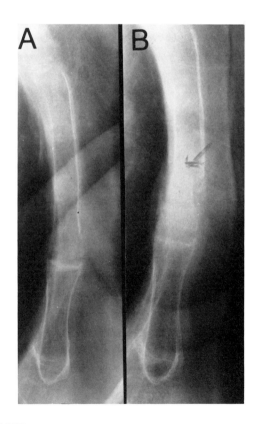

FIG 249.
A, radiograph of the sternum in an osteoporotic woman obtained 1 year before fracture was identified. **B,** 1 year later, a transverse fracture through the upper body of the sternum is seen (*arrow*). (From Cooper KL: Insufficiency fractures of the sternum: A consequence of thoracic kyphosis? *Radiology* 1988; 167:471. Reproduced by permission).

also be a transverse component across the body of the sacrum. If the fracture becomes complete, displacement may occur. On nuclear scanning there is often a characteristic H-shaped or butterfly appearance of the uptake across the sacrum and sacroiliac joints in cases involving fractures of both sacral alae and the body of the sacrum.[361, 362]

Plain-film findings are difficult to visualize and are usually seen as a longitudinal band of sclerosis. Recognition is rendered difficult by overlying bowel shadow. A distinct fracture line can be more easily identified in those cases in which there is disruption of the cortical margin of one of the anterior sacral foramina or when bilateral sacral fractures are joined by a horizontal fracture on the lateral radiograph of the sacrum.[359] Figure 250 illustrates such a fracture of the sacrum, discernible by plain-film radiography. Figure 251 illustrates the more common situation where the fracture was not discernible by plain film and required nuclear scanning and CT to demonstrate the lesion.

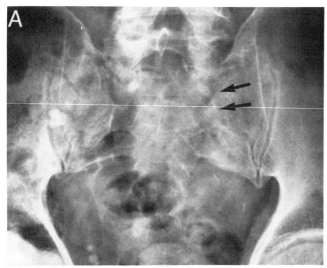

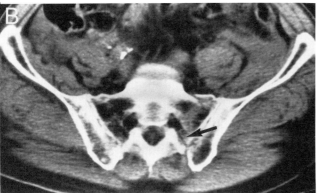

FIG 250.
A, insufficiency fracture of the left sacral ala seen as a vertical band of sclerosis with disruption of the cortical margins of the anterior sacral foramina (*arrows*). **B,** computed tomographic scan demonstrates the fracture to good advantage (*arrow*).

THE ILIUM

The most common stress fracture seen in the ilium occurs in the supra-acetabular region. This fracture occurs as the result of postmenopausal osteoporosis, steroid therapy, radiotherapy, and rheumatoid arthritis. These fractures are seen as hazy bands of sclerosis located immediately above and parallel to the acetabular roof and result in hip pain (Fig 252). Many of these patients have additional fractures in the spine or pelvis.[34]

Stress fracture of the pelvis is not uncommon in patients with rheumatoid arthritis and has been described as occurring through the wing of the ilium, the acetabulum, and the superior and inferior pubic rami.[364] Acetabular stress fractures have also been described in an osteoporotic paraplegic patient as a result of ambulation therapy.[365] A stress fracture of the iliac bone following removal of a large iliac graft has also been reported.[366]

THE PUBIS

Parasymphyseal Lesions

Postmenopausal women and women with rheumatoid arthritis are subject to insufficiency fractures of the parasymphyseal area of the os pubis. These fractures closely simulate the radiographic changes of malignant neoplasia. Some of these patients will describe a history of trauma or recent increase in activity.[363, 367–370] These patients complain of groin pain, which may follow a period of increased activity. Radiographic examination shows fractures that extend parallel and adjacent to the symphysis pubis, or, on occasion, the fracture may be horizontal through the flat junction of the superior and inferior pubic rami. Motion of the fracture will result in areas of lysis and callus formation, which produce the appearance of a destructive malignant-appearing lesion (Figs 253 through 255).

The Rami

The site of closure of the ischiopubic junction is a common location for stress fractures in the healthy skeleton. It is also a frequent location for insufficiency fractures. Figure 256 shows an insufficiency fracture of the pubis in a 50-year-old woman that followed 2 days of lifting boxes. The proximal portion of the pubis adjacent to the acetabulum is also a common locus of fracture (Fig 257). Pubic ramus fractures may also occur in conjunction with parasymphyseal fractures (Fig 258). Stress fractures of the inferior pubic rami following hip surgery are described, probably related to the stress of altered weight bearing.[371–373] Insufficiency stress fractures associated with degenerative arthritis of the hip[374] or developing after total knee replacement and arthroplasty,[375, 376] and after tibial osteotomy,[377] have also been reported.

THE FEMUR

The Femoral Neck

Fractures of the femoral neck in osteoporotic women are second in incidence only to fractured vertebrae and are a significant threat to the well-being of the osteoporotic woman. These fractures are usually induced by trauma and are not, therefore, stress fractures in the usual sense. However, true stress fractures of the femoral neck have been described in patients with rheumatoid arthritis.[378–380] Stress fractures in this location have also been seen as a complication of internal fixation of a traumatic fracture,[381] after total replacement of the knee,[382, 383] and following total hip arthroplasty.[384]

The Femoral Shaft

It is not commonly appreciated that women with severe osteoporosis may develop bowing of their femurs and suffer stress fractures on the convex side of the shaft as a

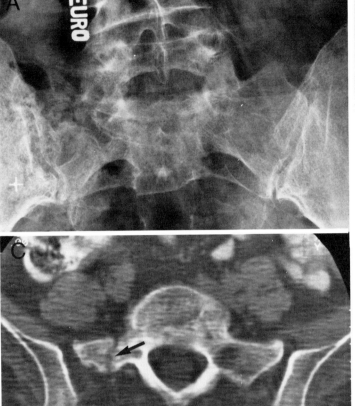

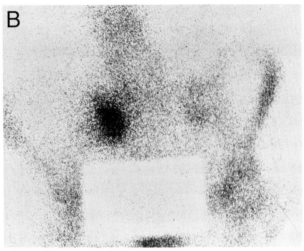

FIG 251.
Insufficiency fracture of the right sacral ala in a 68-year-old woman. **A,** plain film does not define the fracture. **B,** nuclear scan shows increased uptake in the right sacral ala. **C,** computed tomographic scan demonstrates a vertical fracture of the right ala (*arrow*).

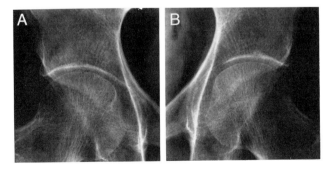

FIG 252.
A 57-year-old woman with osteoporosis and pain in her right hip. **A,** ill-defined band of sclerosis arches above roof of right acetabulum. **B,** opposite, normal hip for comparison. (From Cooper KL, Beabout JW, McLeod RA: Supra-acetabular insufficiency fractures. *Radiology* 1985; 157:15. Reproduced by permission.)

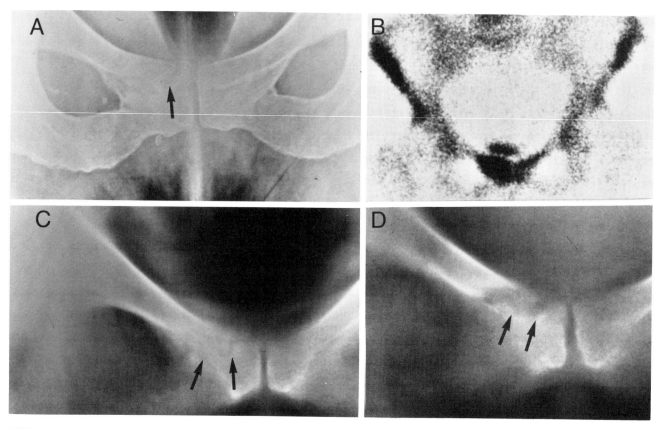

FIG 253.
Parasymphyseal insufficiency fracture in a 60-year-old osteoporotic woman. **A,** plain film shows fracture line (*arrow*). **B,** bone scan shows increased activity in right parasymphyseal area. **C** and **D,** tomograms show horizontal fracture (*arrows*).

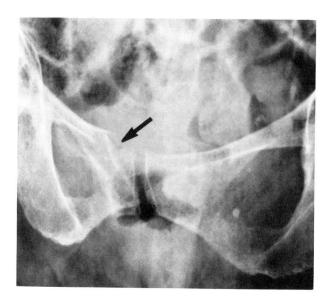

FIG 254.
Parasymphyseal insufficiency fracture in a 79-year-old osteoporotic woman (*arrow*).

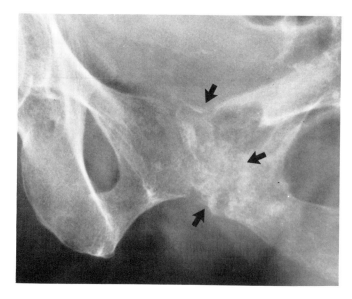

FIG 255.
Parasymphyseal insufficiency fracture in a 59-year-old osteoporotic woman with breast cancer, mistaken for a metastatic deposit (*arrows*).

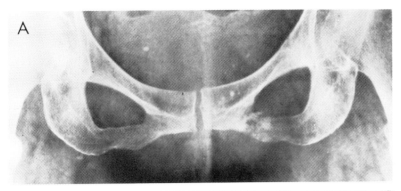

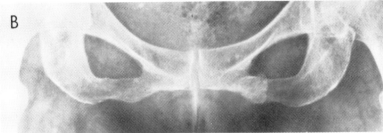

FIG 256.
A, insufficiency fracture of the left pelvis in a 50-year-old osteoporotic woman. B, follow-up film obtained 2 months later shows healing.

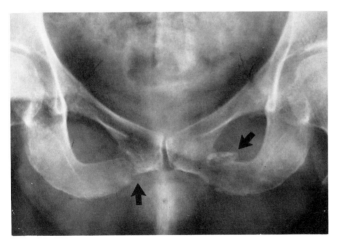

FIG 257.
Insufficiency fracture of the superior and inferior pubic rami in a 70-year-old osteoporotic woman (*arrows*).

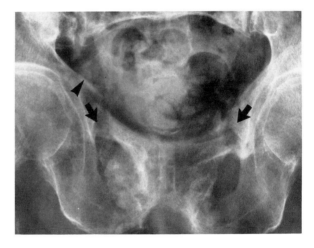

FIG 258.
Insufficiency fractures of the superior rami of the pubis associated with parasymphyseal fractures in a 74-year-old osteoporotic woman (*arrows*).

result of weight bearing. Many of these cases are described in the literature as examples of osteomalacia with Looser's transformation zones. Such a case is illustrated in Figure 259. This 82-year-old woman had severe osteoporosis but no evidence of osteomalacia. Her femora were bowed laterally and showed multiple cortical stress fractures similar to those seen in the bowed long bones of patients with Paget's disease. Similar osteoporotic stress fractures simulating pseudofractures of osteomalacia have also been de-

scribed in other locations, including the femoral neck,[385] the femora and pubic rami,[386] and the femur and metatarsals.[387]

Metaphyseal impaction insufficiency fractures of the distal metaphyses of the femur, tibia, and fibula have been described by Manson et al.[388] in children with acute lymphatic leukemia. These children had severe osteoporosis and were seen in some cases before chemotherapy was administered.

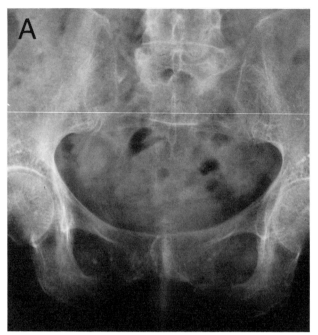

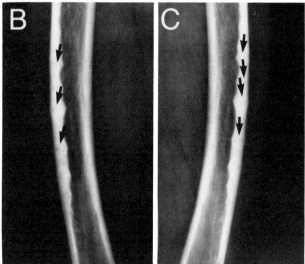

FIG 259.
A to **C,** severe osteoporosis in an 82-year-old woman with lateral bowing of her femora and cortical stress fractures on the lateral aspect of the femora (*arrows*).

THE PATELLA

Patellar stress fractures are usually seen in older patients after patellar resurfacing and more often in patients with degenerative arthritis.[389] They are also likely to be seen in patients with severe osteoporosis, including those with rheumatoid arthritis. Figure 260 shows an insufficiency fracture of the patella in a 56-year-old osteoporotic paraplegic woman.

THE TIBIA

The tibial plateau is one of the more common sites of occurrence of insufficiency fracture.[390, 391] The diagnosis is often not suspected until a bone scan is performed and characteristically shows an area of intense increase in activity in the area of the fracture. On radiographs the fracture is usually seen as a horizontal, linear band of sclerosis. A lucent fracture line or periosteal proliferation is not usually present (Fig 261). If the fracture is not recognized early in its course, collapse of the trabecular bone at the fracture site may occur with resultant depression of the plateau (Fig 262).

Stress fractures in osteoporotic women have been described in this location or, more distally, in the shaft in association with degenerative arthritis of the knee[392–394] and following total knee arthroplasty[395] and high tibial osteotomy.[396] Figure 263 illustrates such an insufficiency fracture of the distal tibia in an 80-year-old woman following hip replacement on the same side.

Patients with rheumatoid arthritis also develop insufficiency fractures of the tibia in the midshaft of the bone as well as in either metaphysis.[397–400] Midshaft tibial insufficiency fractures have also been discovered in elderly patients with pyrophosphate arthropathy[401] and in patients receiving fluoride therapy for osteoporosis.[402]

Insufficiency fractures of the distal end of the tibia may be quite subtle and evidenced by minimal sclerosis and/or periosteal reaction. The area of increased nuclear uptake in some of these fractures may be quite large and may mislead the observer to suspect the presence of a neoplasm (Fig 264). Although most stress fractures of the tibia are transverse or oblique, insufficiency fractures may also be longitudinal and may be responsible for such large areas of increased nuclear uptake[20, 403] (Figs 265 and 266).

History is often not contributory to the diagnosis. Figure 267 shows insufficiency fractures of the tibia and fibula in a 60-year-old woman resulting from climbing stairs to a new apartment over a period of several days. Figure 268 illustrates the case of a 52-year-old woman whose roentgenographic examination shows a healing stress fracture that was misinterpreted since there was no history of trauma.

Insufficiency fractures of the tibia are often associated with similar fractures of the fibula at the same level. The fibular fracture may show displacement if immobilization is delayed (Fig 269).

The Calcaneus

Calcaneal insufficiency fractures may be seen in patients with postmenopausal or senile osteoporosis and present identically to those in the normal skeleton (Figs 270 and 271). These fractures have also been described in patients under treatment with fluoride for osteoporosis[404, 405] and in an elderly woman with pancreatitis and subcutaneous fat necrosis.[406] In addition, spontaneous rupture of the tendon of Achilles has been described in patients with rheumatoid arthritis.[407]

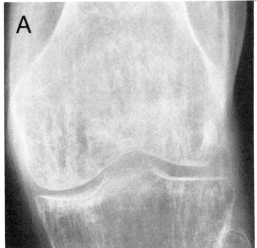
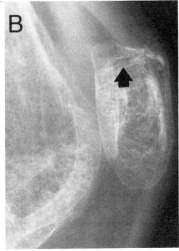

FIG 260.
Insufficiency fracture of the patella (*arrow*) in a 56-year-old paraplegic woman with severe osteoporosis. **A,** frontal projection. **B,** lateral projection.

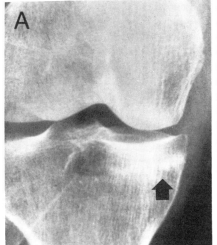

FIG 261.
Tibial plateau insufficiency fracture in a 68-year-old woman. **A,** plain film shows band of sclerosis at medial aspect of the tibial plateau (*arrows*). **B,** nuclear scan shows localized increased uptake in the area of the fracture.

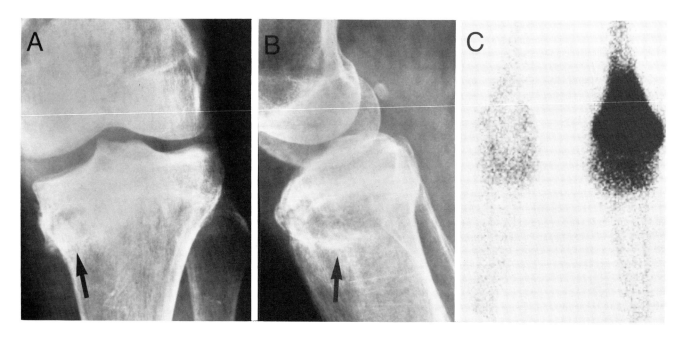

FIG 262.
Insufficiency fracture of the medial plateau (*arrows*) in a 55-year-old osteoporotic woman with collapse of the metaphyseal bone. **A** and **B**, plain films. **C**, nuclear scan.

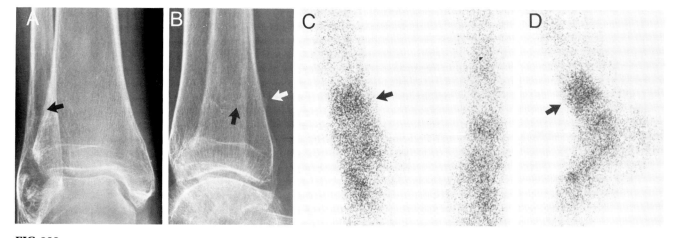

FIG 263.
Insufficiency fracture of the distal tibia in an 80-year-old woman following ipsilateral hip replacement (*arrows*). **A** and **B**, plain films. Fracture is reflected in minimal periostitis and sclerosis. **C** and **D**, nuclear scans.

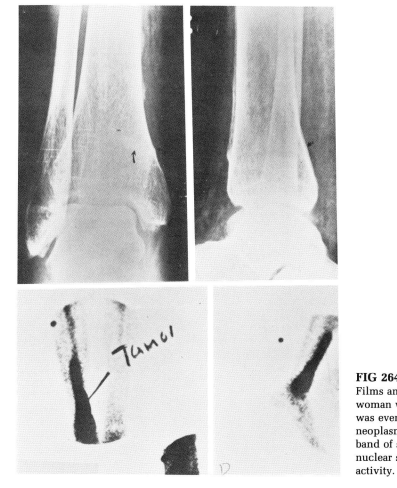

FIG 264.
Films and scans of the tibia of a 55-year-old woman with an insufficiency fracture (*arrows*) that was eventually subjected to biopsy for suspected neoplasm. **A** and **B,** plain films show minimal band of sclerosis and periosteal reaction. **C** and **D,** nuclear scans show very large area of increased activity.

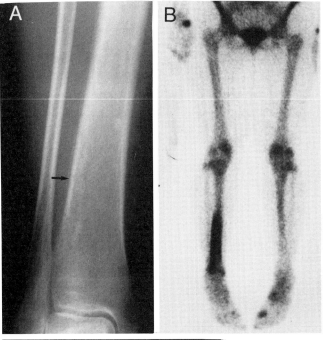

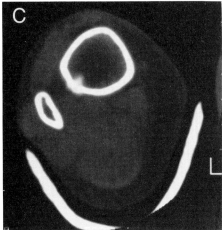

FIG 265.
A 60-year-old osteoporotic woman with longitudinal stress
fracture of the tibia. **A,** plain radiograph demonstrates periosteal
reaction (*arrow*) along the course of the interosseous membrane.
No fracture line is apparent. **B,** anteroposterior bone scan shows
increased uptake in the right tibia, corresponding to **A.** The left
medial foot is affected by focal arthritis. The tibia was normal 7
months later. **C,** computed tomographic scan demonstrates the
lucent fracture line as well as the periosteal and endosteal
callus. The fracture was apparent on 16 contiguous 5 mm–thick
sections. (From Allen GJ: Longitudinal stress fractures of the
tibia: Diagnosis with CT. *Radiology* 1988; 167:799. Reproduced
by permission.)

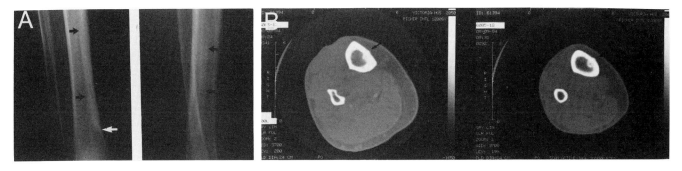

FIG 266.
Longitudinal stress fracture of the tibia in a 59-year-old woman. **A,** anteroposterior and lateral radiographs of the right tibia reveal a 17-cm vertical band of increased density in the shaft of the tibia (*black arrows*) and a focal cortical interruption inferiorly with periosteal reaction (*white arrow*). **B,** computed tomographic scans of the tibia at two levels demonstrate the vertical fracture line with periosteal and endosteal reactive bone (*arrows*). (From Miniaci A, McLaren AC, Haddad RG: Longitudinal stress fracture of the tibia: Case report. *J Can Assoc Radiol* 1988; 39:221–223. Reproduced by permission.)

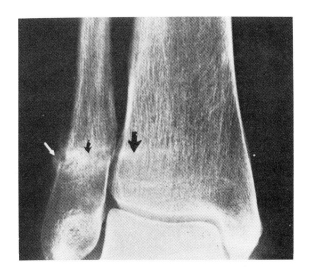

FIG 267.
Insufficiency fractures of the tibia and fibula (*arrows*) in a 60-year-old woman resulting from stair climbing.

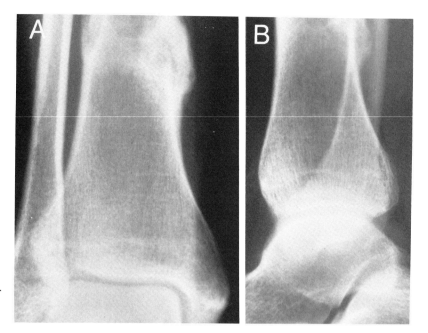

FIG 268.
Healing insufficiency fracture of the tibia in a 52-year-old woman. **A,** frontal projection. **B,** lateral projection.

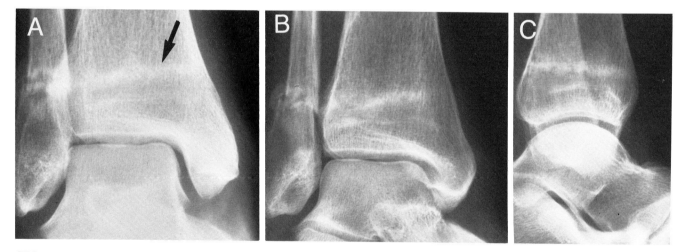

FIG 269.
Insufficiency fractures of the tibia (*arrow*) and fibula in a 59-year-old woman. Note displacement of the distal fibula segment in **B. A,** anteroposterior projection. **B,** oblique projection. **C,** lateral projection.

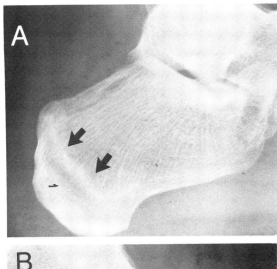

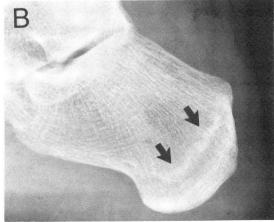

FIG 270.
Bilateral asymmetric calcaneal insufficiency fractures (*arrows*) in a 64-year-old osteoporotic woman. **A,** left foot. **B,** right foot.

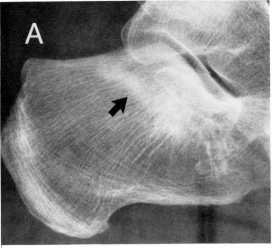

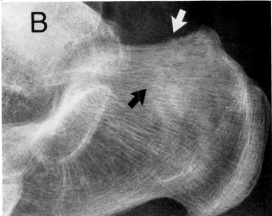

FIG 271.
A and **B,** bilateral asymmetric calcaneal stress fractures (*arrows*) in a 64-year-old woman secondary to prolonged standing. In **B,** note disruption of the superior cortex.

9

Osteomalacia

The radiologic changes of osteomalacia are well documented. Deformities of the long bones, spine, pelvis, and skull due to bone softening are characteristic. Spongy bone shows a decreased number of trabeculae, and those that remain appear prominent and coarse. Pseudofractures or Looser's zones are characteristic of this condition.[408–410] These lucent areas are oriented at right angles to the cortex and incompletely span the bone. The sites tend to be bilateral and symmetric. Looser's zones tend to occur in characteristic sites such as the axillary margins of the scapula, ribs (Fig 272), superior and inferior pubic rami (Fig 273), inner margins of the proximal femora, and posterior margins of the proximal ulnae. In the femora it is noteworthy that they occur on the medial side of the bone as opposed to stress fractures secondary to bowing, which occur on the lateral convex aspect of the bone.

Insufficiency stress fractures also occur in osteomalacia, usually in weight-bearing bones, in contrast to the pseudofractures, which do not necessarily involve weight-bearing bones (Figs 274 through 276).

The etiology of pseudofractures or Looser's zones in osteomalacia is not clear. It has been suggested that they are secondary to mechanical erosion by adjacent blood vessels or to sites of stress and accelerated bone turnover.

Insufficiency fractures have also been described in osteomalacia associated with renal failure, in the ilium,[411] in the femoral neck,[412] and in the tibia and fibula.[413]

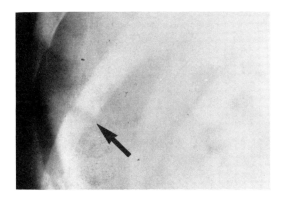

FIG 272.
Pseudofracture of the ribs in osteomalacia (*arrows*).

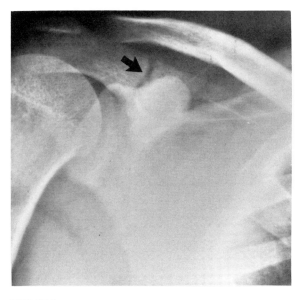

FIG 274.
Insufficiency fractures of the acromion in a patient with osteomalacia (*arrow*).

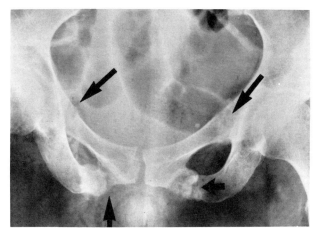

FIG 273.
Pseudofractures of the superior and inferior pubic rami in osteomalacia (*arrows*).

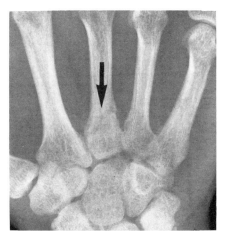

FIG 275.
Insufficiency fracture of the third metacarpal in a patient with osteomalacia (*arrow*).

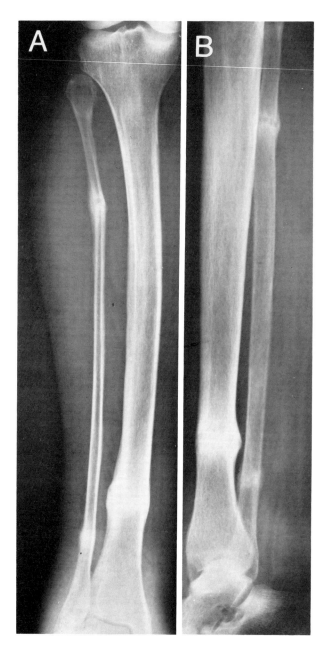

FIG 276.
Insufficiency fractures of the tibia and fibula in a patient with
osteomalacia. Note bowing of the shafts of the tibia and fibula.
A, frontal projection. **B,** lateral projection.

10

Paget's Disease

Long-standing Paget's disease leads to bowing deformities of the femora and tibiae. On the lateral, convex side of the bowed femur, one may see multiple horizontal lucencies in the cortex, which represent stress fractures.[414, 415] These are identical in appearance, location, and etiology to the stress fractures seen in the laterally bowed femora of osteoporotic women (Fig 277). These insufficiency stress fractures should not be mistaken for the pseudofractures (Looser's zones) of osteomalacia.

Cortical stress fractures of this type are seen in the femur and the tibia and may appear as a single cortical cleft or at multiple levels. In the tibia, they occur in the anterior cortex since the bone usually bows anteriorly. As in the femur, these insufficiency stress fractures should not be mistaken for pseudofractures (Looser's zones) of osteomalacia, which involve the concave side of the bone rather than the convex side of the bone as in Paget's disease.

In Paget's disease, these clefts are filled with cartilaginous callus that does not mineralize adequately and is therefore radiolucent.[416] These fractures may partially heal on their periosteal and endosteal surfaces, isolating a permanent central radiolucent area[417] (see Fig 277). In other patients, the partial fracture does not heal or may progress across the entire bone, thus representing pathologic fracture.

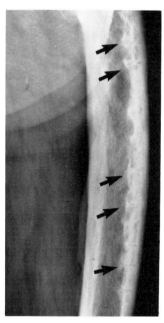

FIG 277.
Paget's disease with bowed femur and multiple stress fractures of the lateral cortex (*arrows*).

PART III

Overuse Syndromes

11

Muscle Overuse Syndromes

Muscle damage resulting from extreme exercise can be demonstrated by Tc-99m pyrophosphate scintigraphy and is evidenced by localized increased muscle radionuclide concentrations. The detection of muscle damage appears to be best when the examination is performed within 48 hours after the exercise.[419] Figure 278 shows such an example in a young man with acute pain and swelling over his lumbar spine after prolonged exercise. It is debatable whether such changes represent rhabdomyolysis, in which case myoglobulin should be present in the urine. Rhabdomyolysis is usually secondary to compression of muscle, trauma, metabolic derangements, drug abuse, and hypothermia and may lead to renal failure.[420] In confirmed cases of rhabdomyolysis, nuclear scanning will show an increase of uptake of nuclide in the affected muscles.[421] Computed tomography has also been useful in showing swelling of the involved muscles in rhabdomyolysis[420] (Fig 279). Muscle swelling can also be demonstrated by ultrasound as an area of decreased echogenicity.[422]

Another interesting by-product of muscle overexertion is illustrated in Figure 280, which demonstrates spontaneous pneumomediastinum with extension into the cervical soft tissues in a 21-year-old weight lifter who experienced crepitus in the neck during lifting. This injury followed straining against a closed glottis with resultant increase in intrathoracic pressure.

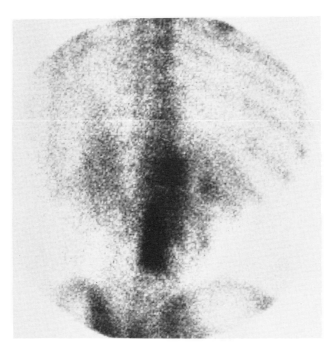

FIG 278.
Acute muscle injury of the paraspinous muscles of a 29-year-old man after prolonged lifting. Note localized increased uptake in the lumbar area.

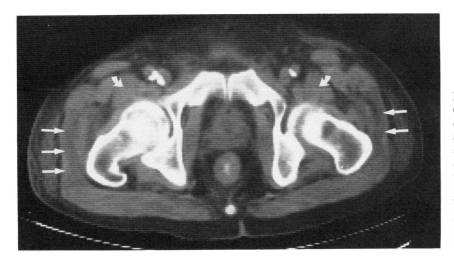

FIG 279.
Computed tomographic scan at the level of the acetabular fossa reveals patchy areas of low attenuation with minimal swelling involving both gluteus muscles along the flanks (*straight arrows*). Note the normal attenuation of the iliopsoas muscles (*curved arrows*). (From Barloon TJ, Zachar CK, Harkens KL, et al: Rhabdomyolysis: Computed tomography findings. CT: *J Comput Tomogr* 1988; 12:193. Reproduced by permission).

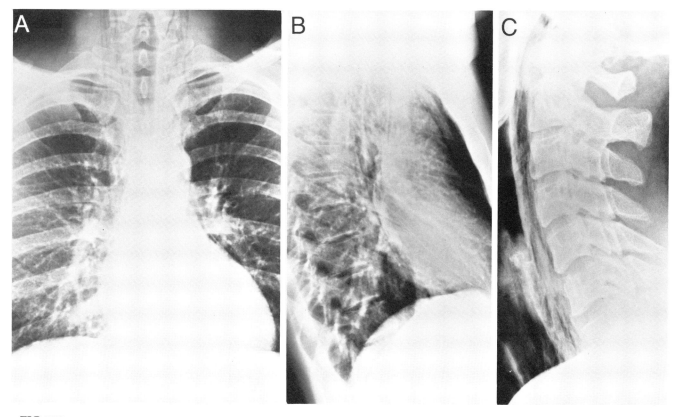

FIG 280.
Spontaneous pneumomediastinum and cervical interstitial emphysema in a 21-year-old weight lifter. **A,** frontal projection. **B,** lateral projection. **C,** lateral projection of the soft tissues of the neck.

12

Skeletal Overuse Syndromes

Overuse of joints in athletics or in occupations will eventuate in isolated or premature degeneration of the stressed joint. It is interesting that certain individuals are more prone to this phenomenon than others involved in the same occupation or sport, a circumstance probably related to the anatomy and structure of their joints and cartilage. These overuse activities lead to atypical degenerative joint disease, often in joints that do not usually degenerate selectively. Figure 281 shows degenerative arthritis of the right shoulder in a 70-year-old man who has participated vigorously in athletics all his adult life, with overuse of his right shoulder in tennis, football, and basketball. Figure 282 shows similar changes in both shoulders of a 55-year-old white-water canoer. Similar shoulder changes have been reported in baseball pitchers,[97] weight lifters, and polo players.[423]

Figure 283 shows premature degeneration of the acromioclavicular joint in a 40-year-old circus performer whose act involved supporting a 100-lb dwarf overhead with his right arm. Similar acromioclavicular and sternoclavicular joint degeneration has been reported after excessive muscular stretching.[103]

Premature degeneration of the elbow joints has been implicated in a variety of sports and occupations. Figure 284 shows marked degeneration of both elbows in a 31-year-old weight lifter who self-administered intra-articular corticosteroids every week for a year to reduce the pain in his elbows induced by weight lifting. Similar findings without the use of steroids are illustrated in Figure 285 in a 40-year-old weight lifter. Degenerative changes of this type have also been seen in baseball pitchers,[97] handball players,[423] foundry workers,[424] tennis players (Fig 286), and laundry linen pullers (Fig 287).

Premature sclerotic degenerative arthritis of the wrist has been described in polo players[423] and particularly in certain occupations in which there is chronic repetitive trauma. Figure 288 shows degenerative arthritis of the right wrist in a right-handed roofer. Figure 289 shows cystic changes in the lunates in a woman whose work in the shoe making industry required repeated pounding with her wrists. Figure 290 shows bilateral degenerative arthritis of the wrists in a 57-year-old pneumatic drill operator. These changes have been reported not only in the wrists of pneumatic drill operators[425–427] but also in operators of chain saws[428] (see Fig 271).

Other associations of degenerative arthritis with sports or athletics include degenerative changes in the ankles and feet in ballet dancers and soccer players[429, 430]; the ankles, feet, hips, and knees in lacrosse and soccer players[431–433]; the knees in football players[434]; the patellofemoral joints in cyclists[435]; the hips in farmers[436]; the elbows and knees in coal miners[437]; and the wrists, elbows, and knees in boxers and wrestlers.[438, 439] Workers who do heavy work demanding sustained gripping motions of both hands develop degenerative changes in their metacarpophalangeal joints,[425] as do jackhammer operators.[344] In cotton mill workers[440] and cricket players,[441] the interphalangeal joints are involved.

An arthropathy described in piano players consists of degenerative arthritis of the metacarpophalangeal and distal interphalangeal joints. This is associated with axial rotation of the third through fifth digits with sclerosis and flattening of the phalangeal tufts.[442]

Elliott and Elliott[443] reported a bilateral polyarthritis of the wrists in lumbermen and stone masons whose jobs required lifting. The overstressed ligaments eventually fail and anterior lunate subluxation may result. Figure 291 shows degenerative changes in the talonavicular joints in a 38-year-old long-term soccer player, representing another example of the effects of overuse.

162

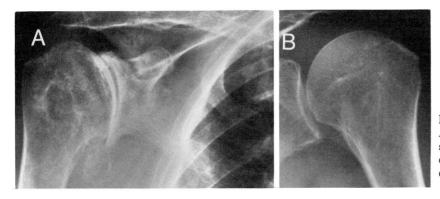

FIG 281.
A, degenerative arthritis of the right shoulder in a 70-year old man secondary to overuse in athletics. **B,** left shoulder for comparison.

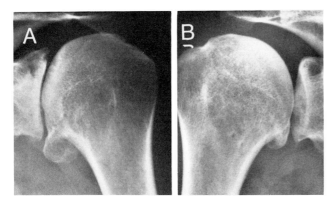

FIG 282.
A and **B,** degenerative arthritis of both shoulders in a 55-year-old whitewater canoer.

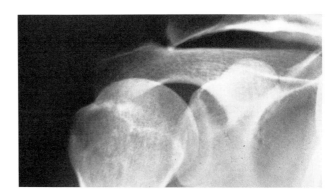

FIG 283.
Premature degenerative arthritis of the acromioclavicular joint in a 40-year-old circus performer whose act required prolonged lifting with the left arm.

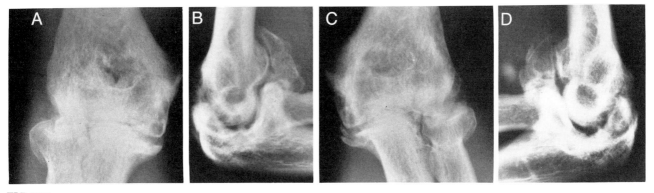

FIG 284.
Bilateral degenerative arthritis of the elbows in a 31-year-old weight lifter who self-administered interarticular steroids for a year. **A** and **B,** right elbow. **C** and **D,** left elbow.

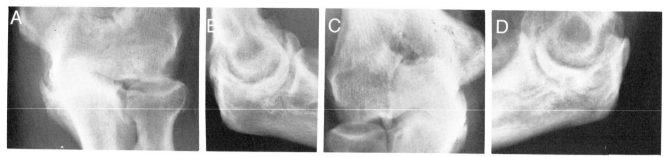

FIG 285.
Degenerative arthritis of the elbows in a 40-year-old long-term weight lifter. **A** and **B**, left elbow. **C** and **D**, right elbow.

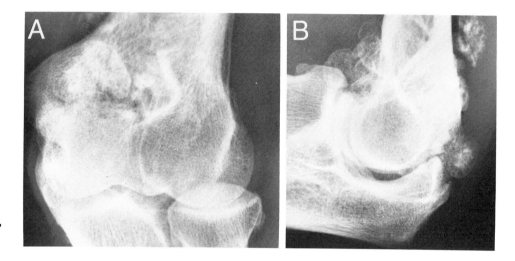

FIG 286.
Degenerative arthritis of the elbow in a 59-year-old tennis player. **A**, oblique projection. **B**, lateral projection.

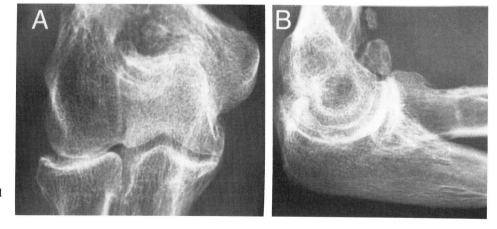

FIG 287.
Degenerative arthritis of the right elbow in a right-handed 60-year-old laundry linen puller. **A**, frontal projection. **B**, lateral projection.

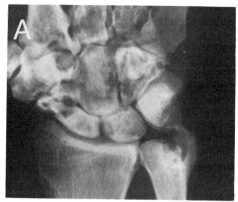
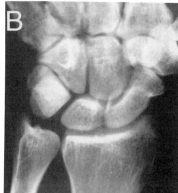

FIG 288.
Degenerative arthritis of the right wrist in a right-handed roofer who had worked at this occupation for 30 years. **A,** right wrist, **B,** left wrist.

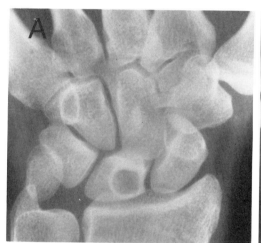
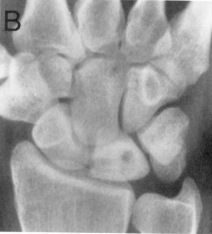

FIG 289.
A and **B,** a 40-year-old woman employed in shoe making with cystic changes in her lunates secondary to trauma sustained by repetitive pounding with her wrists.

CONCLUSION

A reading of the data presented in this text will confirm the protean nature of the radiologic manifestations of musculoskeletal stress injury. The necessity for prompt and accurate diagnosis is emphasized by the wide variety of misdiagnoses which can and are made by the unwary who do not consider musculoskeletal stress as the inciting insult. The literature provides an ever-expanding account of new locations and radiologic appearances of stress injury. These have become so commonplace in practice that one must always include in one's differential thinking the possibility that the lesion at hand is the result of musculoskeletal stress in either the normal or abnormal skeleton. Awareness is the key to prompt and correct diagnosis.

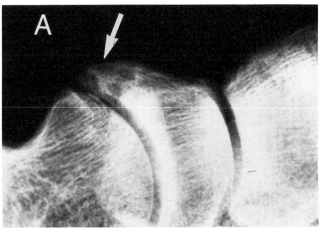

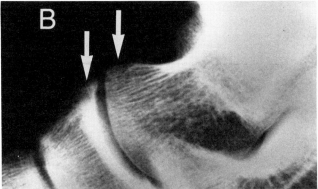

FIG 291.
Degenerative changes in the talonavicular joints in a 38-year-old soccer player manifested by hypertrophic lipping of the joint margins (*arrows*). **A,** left foot. **B,** right foot.

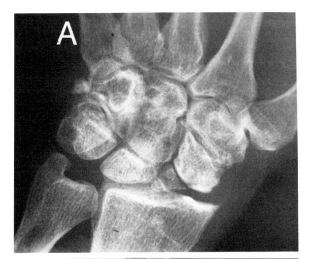

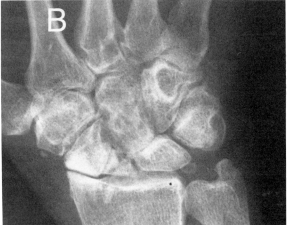

FIG 290.
A and **B,** bilateral degenerative arthritis of the wrists in a 57-year-old pneumatic drill operator.

References

1. Breithaupt MD: Zur Pathologie des menschlichen Fussen. *Med Zeitung* 1855; 24:169–171, 175–177.
2. Pauzet TE: De la periostite osteoplastique des metatarsiens a la suite des marches. *Arch Med Pharm Mil* 1887; 10:337–353.
3. Stechow AW: Fussoedem und Rontgenstrahlen. *Dtsch Mil-Aerztl Z* 1897; 26:465–471.
4. Blecher A: Uber den Einfluss der Parademarche auf die Entstehung der Fussgeschwuste. *Med Klin* 1905; 1:305–306.
5. Deutschlander C: Uber entzundliche mittel fuss-geschwutse. *Arch Klin Chir* 1921; 118:530–544.
6. Asal W: Uberlastungsschaden am Knochensystem bei Soldaten. *Arch Klin Chir* 1936; 186:511–522.
7. Morris JM, Blickenstaff LO: *Fatigue Fractures: A Clinical Study.* Springfield, Ill, Charles C Thomas Publishers, 1967.
8. Devas MB: *Stress Fractures.* Edinburgh, Churchill Livingstone, Inc, 1975.
9. Johnson LC, Stradford HT, Geis RW, et al: Histiogenesis of stress fractures. *J Bone Joint Surg [Am]* 1963; 45:1542.
10. Markey, KL: Stress fractures. *Clin Sports Med* 1987; 6:405–425.
11. Wilson ES Jr, Katz FN: Stress fractures: An analysis of 250 consecutive cases. *Radiology* 1969; 92:481–486.
12. Greaney RB, Gerber FH, Laughlin RL, et al: Distribution and natural history of stress fractures in U.S. Marine recruits. *Radiology* 1983; 146:339–346.
13. Matheson GO, Clement DB, McKenzie DC, et al: Stress fractures in athletes: A study of 320 cases. *Am J Sports Med* 1987; 15:46–58.
14. McBryde AM: Stress fractures in athletes. *J Sports Med* 1975; 3:212–217.
15. Orava S, Puranen J, Ala-Ketola L: Stress fractures caused by physical exercise. *Acta Orthop Scand* 1978; 49:19–27.
16. Maffulli N, Regine R, Angelillo M, et al: Ultrasound diagnosis of Achilles tendon pathology in runners. *Br J Sports Med* 1987; 21:158–162.
17. Savoca CJ: Stress fractures: A classification of the earliest radiographic signs. *Radiology* 1971; 100:519–524.
18. Uhthoff HK, Jaworski ZFG: Periosteal stress-induced reactions resembling stress fractures: A radiologic and histologic study in dogs. *Clin Orthop* 1985; 199:284–291.
19. Floyd WN Jr, Butler JE, Clanton T, et al: Roentgenologic diagnosis of stress fractures and stress reactions. *South Med J* 1987; 80:433–439.
20. Allen GJ: Longitudinal stress fractures of the tibia: Diagnosis with CT. *Radiology* 1988; 167:799–801.
21. Rupani HD, Holder LE, Espinola DA, et al: Three-phase radionuclide bone imaging in sports medicine. *Radiology* 1985; 156:187–196.
22. Holder LE, Matthews LS: The nuclear physician and sports medicine, in Freeman LM, Weissmann HS (eds): *Nuclear Medicine Annual: 1984.* New York, Raven Press, 1984, pp 81–140.
23. Matin P: The bone scan in traumatic and sports injuries, in Fogelman I (ed): *Bone Scanning in Clinical Practice.* New York, Springer-Verlag, 1987, pp 121–132.
24. Matin P: The appearance of bone scans following fractures, including immediate and long-term studies. *J Nucl Med* 1979; 20:1227–1231.
25. Zwas ST, Elkanovitch R, Frank G: Interpretation and classification of bone scintigraphic findings in stress fractures. *J Nucl Med* 1987; 28:452–457.
26. Meurman KOA, Elfving S: Stress fracture in soldiers: A multifocal bone disorder. *Radiology* 1980; 134:483–487.
27. Ben Ami T, Treves ST, Tumeh S, et al: Stress fractures after surgery for osteosarcoma: Scintigraphic assessment. *Radiology* 1987; 163:157–162.
28. Prather JL, Nusynowitz ML, Snowdy HA, et al: Scintigraphic findings in stress fractures. *J Bone Joint Surg [Am]* 1977; 59:869–874.
29. Somer K, Meurman KOA: Computed tomography of stress fractures. *J Comput Assist Tomogr* 1982; 6:109–115.
30. Murcia M, Brennan RE, Edeiken J: Computed tomography of stress fracture. *Skeletal Radiol* 1982; 8:193–195.
31. Yousem D, Magid D, Fishman EK, et al: Computed tomography of stress fractures. *J Comput Assist Tomogr* 1986; 10:92–95.
32. Cooper KL, Beabout JW, Swee RG: Insufficiency fractures of the sacrum. *Radiology* 1985; 156:15–20.
33. Gacetta DJ, Yandow DR: Computed tomography of spontaneous osteoporotic sacral fractures. *J Comput Assist Tomogr* 1984; 8:1190–1191.
34. Cooper KL, Beabout JW, McLeod RA: Supra-acetabular insufficiency fractures. *Radiology* 1985; 157:15–17.
35. Rawlings CE III, Wilkins RH, Martinez S, et al: Osteoporotic sacral fractures: A clinical study. *Neurosurgery* 1988; 22:72–76.
36. Stafford SA, Rosenthal DI, Gebhardt MC, et al: MRI in stress fracture. *AJR* 1986; 147:553–556.
37. Yao L, Lee JK: Occult intraosseous fracture: Detection with MR imaging. *Radiology* 1988; 167:749–751.
38. Lee JK, Yao L: Stress fractures: MR imaging. *Radiology* 1988; 169:217–220.
39. Brunner MC, Flower SP, Evancho AM, et al: MRI of the athletic knee: Findings in asymptomatic professional basketball and collegiate football players. *Invest Radiol* 1989; 24:72–75.
40. Devereaux MD, Parr GR, Lachmann SM, et al: The diagnosis of stress fracture in athletes. *JAMA* 1984; 252:531–533.
41. Giladi M, Ziv Y, Aharonson Z, et al: Comparison between radiography, bone scan, and ultrasound in the diagnosis of stress fractures. *Milit Med* 1984; 149:459–461.
42. Giladi M, Milgrom C, Danon Y, et al: The correlation between cumulative march training and stress fractures in soldiers. *Milit Med* 1985; 150:600–601.
43. Dotter WE: Little Leaguer's shoulder: A fracture of

the proximal epiphyseal cartilage of the humerus due to baseball pitching. *Guthrie Clin Bull* 1953; 23:68–72.

44. Adams JE: Little League shoulder: Osteochondrosis of the proximal humeral epiphysis in boy baseball pitchers. *Calif Med* 1966; 105:22–25.
45. Hansen NM: Epiphyseal changes in the proximal humerus of an adolescent baseball pitcher. *Am J Sports Med* 1982; 10:380–384.
46. Allen ME: Stress fracture of the humerus: A case study. *Am J Sports Med* 1984; 12:244–245.
47. Chao SL, Miller M, Teng SW: A mechanism of spiral fracture of the humerus: A report of 259 cases following the throwing of hand grenades. *J Trauma* 1971; 11:602.
48. Waris W: Elbow injuries of javelin-throwers. *Acta Chir Scand* 1947; 93:563–575.
49. Moon MS, Kim I, Han IH, et al: Arm wrestler's injury. *Clin Orthop* 1980; 147:219–221.
50. Santavirta S, Kiviluoto O: Transverse fracture of the humerus in a shotputter: A case report. *Am J Sports Med* 1977; 5:122–123.
51. Ishikawa H, Ueba Y, Yonezawa T, et al: Osteochondritis dissecans of the shoulder in a tennis player. *Am J Sports Med* 1988; 16:547–550.
52. Gore RM, Rogers LF, Bowerman J, et al: Osseous manifestations of elbow stress associated with sports activities. *AJR* 1980; 134:971–977.
53. Retrum RK, Wepfer JF, Olen DW, et al: Case report 355. *Skeletal Radiol* 1986; 15:185–187.
54. Torg JS, Moyer RA: Non-union of a stress fracture through the olecranon epiphyseal plate observed in an adolescent baseball pitcher: A case report. *J Bone Joint Surg [Am]* 1977; 59:264–265.
54a. Brower AC, Neff JR, Tillema DA: An unusual scapular stress fracture. *AJR* 1977; 129:519–520.
55. Pavlov H, Torg JS, Jacobs B, et al: Nonunion of olecranon epiphysis: Two cases in adolescent baseball pitchers. *AJR* 1981; 136:819–820.
56. Heyse-Moore GH, Stoker DJ: Avulsion fractures of the scapula. *Skeletal Radiol* 1982; 9:27–32.
57. Pappas AM: Elbow problems associated with baseball during childhood and adolescence. *Clin Orthop* 1982; 164:30–41.
58. Priest JD, Jones HH, Nagel DA: Elbow injuries in highly skilled tennis players. *J Sports Med* 1974; 2:137–149.
59. Gugenheim JJ Jr, Stanley RF, Woods GW, et al: Little League survey: The Houston study. *Am J Sports Med* 1976; 4:189–200.
60. Norwood LA, Shook JA, Andrews JR: Acute medial elbow ruptures. *Am J Sports Med* 1981; 9:16–19.
61. Goldberg MJ: Gymnastic injuries. *Orthop Clin North Am* 1980; 11:717–726.
62. Bowerman JW, McDonnell EJ: Radiology of athletic injuries: Baseball. *Radiology* 1975; 116:611–615.
63. Miller JE: Javelin thrower's elbow. *J Bone Joint Surg [Br]* 1960; 42:788–792.
64. Fliegel CP: Stress related widening of the radial growth plate in adolescents. *Ann Radiol* 1985; 29:374–376.
65. Read MT: Stress fractures of the distal radius in adolescent gymnasts. *Br J Sports Med* 1981; 15:272–276.
66. Roy S, Caine D, Singer KM: Stress changes of the distal radial epiphysis in young gymnasts: A report of twenty-one cases and a review of the literature. *Am J Sports Med* 1985; 13:301–308.
67. Carter SR, Aldridge MJ: Stress injury of the distal radial growth plate. *J Bone Joint Surg [Br]* 1988; 70:834.
68. Carter SR, Aldridge MJ, Fitzgerald R, et al: Stress changes of the wrist in adolescent gymnasts. *Br J Radiol* 1988; 61:109–112.
69. Vender MI, Watson HK: Acquired Madelung-like deformity in a gymnast. *J Hand Surg [Am]* 1988; 13:19–21.
70. Gumbs VL, Segal D, Halligan JB, et al: Bilateral distal radius and ulnar fractures in adolescent weight lifters. *Am J Sports Med* 1982; 10:375–379.
71. Moss GD, Goldman A, Sheinkop M: Case report 219. *Skeletal Radiol* 1982; 9:148–150.
72. Weigl K, Amrami B: Occupational stress fracture in an unusual location: Report of a case in the distal end of the shaft of the radius. *Clin Orthop* 1980; 147:222–224.
73. Orloff AS, Resnick D: Fatigue fracture of the distal part of the radius in a pool player. *Injury* 1986; 17:418–419.
74. Farquharson-Roberts MA, Fulford PC: Stress fracture of the radius. *J Bone Joint Surg [Br]* 1980; 62:194–195.
75. Eisenberg D, Kirchner SG, Green NE: Stress fracture of the distal radius caused by "wheelies." *South Med J* 1986; 79:918–919.
76. Perry CR, Perry HM III, Burdge RE: Stress fracture of the radius following a fracture of the ulna diaphysis. *Clin Orthop* 1984; 187:193–198.
77. McGoldrick F, O'Brien TM: Bilateral stress fractures of the ulna. *Injury* 1988; 19:360–366.
78. Evans DL: Fatigue fracture of the ulna. *J Bone Joint Surg [Br]* 1955; 37:618–621.
79. Bell RH, Hawkins RJ: Stress fracture of the distal ulna. *Clin Orthop* 1986; 209:169–171.
80. Mutoh Y, Mori T, Suzuki Y, et al: Stress fractures of the ulna in athletes. *Am J Sports Med* 1982; 10:365–367.
81. Rettig AC: Stress fracture of the ulna in an adolescent tournament tennis player. *Am J Sports Med* 1983; 11:103–106.
82. Patel M, Irizarry J, Stricevic M: Stress fracture of the ulnar diaphysis: Review of the literature and report of a case. *J Hand Surg [Am]* 1986; 11:443–445.
83. Newberg AH: Case 3. *Radiographics* 1988; 8:239–240.
84. Hamilton HK: Stress fracture of the diaphysis of the ulna in a body builder. *Am J Sports Med* 1984; 12:405–406.
85. Stark HH, Jobe FW, Boyes JH, et al: Fracture of the hook of the hamate in athletes. *J Bone Joint Surg [Am]* 1977; 59:575–582.
86. Almquist EE: Kienbock's disease. *Clin Orthop* 1986; 202:68–78.
87. Israeli A, Engel J, Ganel A: Possible fatigue fracture of the pisiform bone in volleyball players. *Int J Sports Med* 1982; 3:56–57.
88. Destouet JM, Murphy WA: Guitar player acro-os-

teolysis. *Skeletal Radiol* 1981; 6:275–277.

89. Ord RA, Langdon JD: Stress fracture of the clavicle: A rare late complication of radical neck dissection. *J Maxillofac Surg* 1986; 14:281–284.

90. Cummings CW, First R: Stress fracture of the clavicle after a radical neck dissection: Case report. *Plast Reconstr Surg* 1975; 55:366–367.

91. Kaye JJ, Nance EP Jr, Green NE: Fatigue fracture of the medial aspect of the clavicle. *Radiology* 1982; 144:89–90.

92. Kaplan PA, Resnick D: Stress-induced osteolysis of the clavicle. *Radiology* 1986; 158:139–140.

93. Kruger GD, Rock MG, Munro TG: Condensing osteitis of the clavicle. *J Bone Joint Surg [Am]* 1987; 69:550–557.

94. Franquet T, Lecumberri F, Rivas A, et al: Condensing osteitis of the clavicle. *Skeletal Radiol* 1985; 14:184–187.

95. Sandrock AR: Another sports fatigue fracture. *Radiology* 1975; 117:274.

96. Boyer DW Jr: Trapshooter's shoulder: Stress fracture of the coracoid process. *J Bone Joint Surg [Am]* 1975; 57:862.

97. Bennett GE: Shoulder and elbow lesions of the professional baseball pitcher. *JAMA* 1941; 117:510–514.

98. Benton J, Nelson C: Avulsion of the coracoid process in an athlete. *J Bone Joint Surg [Am]* 1971; 53:356–358.

99. De Rosa GP, Kettlekamp DB: Fracture of the coracoid process of the scapula. *J Bone Joint Surg [Am]* 1977; 59:969.

100. Rounds RC: Isolated fracture of the coracoid process. *J Bone Joint Surg [Am]* 1949; 31:662.

101. Kelly JP: Fractures complicating electroconvulsive therapy and chronic epilepsy. *J Bone Joint Surg [Br]* 1954; 36:70.

102. Ramin JE, Veit H: Fracture of scapula during electric shock therapy. *Am J Psychiatry* 1953; 110:153–154.

103. Rask MR, Steinberg LH: Fracture of the acromion caused by muscle forces. *J Bone Joint Surg [Am]* 1978; 59:1146–1147.

104. Williamson DM, Wilson-MacDonald JW: Bilateral avulsion fractures of the cranial margin of the scapula. *J Trauma* 1988; 28:713–714.

105. Ishizuki M, Yamaura I, Isobe Y, et al: Avulsion fracture of the superior border of the scapula. *J Bone Joint Surg [Am]* 1981; 63:820–822.

106. Wright RS, Lipscomb AB: Acute occlusion of the subclavian vein in an athlete: Diagnosis, etiology and surgical management. *J Sports Med* 1975; 2:343–348.

107. Cormier PJ, Matalon TAS, Wolin PM: Quadrilateral space syndrome: A rare cause of shoulder pain. *Radiology* 1988; 167:797–798.

108. Oechsli WR: Rib fracture from cough: Report of 12 cases. *J Thorac Surg* 1936; 5:530–534.

109. Derbes VJ, Haran T: Rib fracture from muscular effort with particular reference to cough. *Surgery* 1954; 35:294–321.

110. Ginsbury M: Spontaneous fracture of the first rib as a complication of status asthmaticus. *Ann Allergy* 1947; 5:488–489.

111. Even-Tov I, Yedwab GA, Persitz E, et al: Stress fracture of ribs in late pregnancy. *Int Surg* 1979; 64:85–87.

112. Long AE: Stress fracture of the ribs associated with pregnancy. *Surg Clin North Am* 1962; 42:909–919.

113. Warmington WT: Four cases of cough fracture. *Ulster Med J* 1966; 35:72–74.

114. Redmond AD: Chest pain due to stress fractures of the first rib, letter. *Injury* 1982; 13:446–447.

115. Rademaker M, Redmond AD, Barber PV: Stress fracture of the first rib. *Thorax* 1983; 38:312–313.

116. Gurtler R, Pavlov H, Torg JS: Stress fracture of the ipsilateral first rib in a pitcher. *Am J Sports Med* 1985; 13:277–279.

117. Lankenner PA, Micheli LJ: Stress fracture of the first rib. *J Bone Joint Surg [Am]* 1985; 67:159–160.

118. Sacchetti AD, Beswick DR, Morse SD: Rebound rib: Stress-induced first rib fracture. *Ann Emerg Med* 1983; 12:177–179.

119. Curtis JA, Libshitz HI, Dalinka MK: Fracture of the first rib as a complication of midline sternotomy. *Radiology* 1975; 115:63–65.

120. Rasad S: Golfer's fractures of the ribs. *AJR* 1974; 120:901–903.

121. Holden DL, Jackson DW: Stress fracture of the ribs in female rowers. *Am J Sports Med* 1985; 13:342–348.

122. Macones AJ Jr, Fisher MS, Locke JL: Stress-related rib and vertebral changes. *Radiology* 1989; 170:117–119.

123. Cancelmo JJ Jr: Clay shoveler's fracture: A helpful diagnostic sign. *AJR* 1972; 115:540–543.

124. Venable JR, Flake RE, Kilian DJ: Stress fracture of the spinous process. *JAMA* 1964; 190:881–885.

125. Marsh CH: Internal fixation for stress fractures of the ankylosed spine. *J R Soc Med* 1985; 78:377–379.

126. Gelman MI, Umber JS: Fractures of the thoracolumbar spine in ankylosing spondylitis. *AJR* 1988; 130:485–491.

127. Resnick D, Niwayama G: Discovertebral destruction in a man with chronic back problems. *Invest Radiol* 1981; 16:89–94.

128. Borkow SE, Kleiger B: Spondylolisthesis in the newborn: A case report. *Clin Orthop* 1971; 81:73–76.

129. Klinghoffer L, Murdock MG: Spondylolysis following trauma: A case report and review of the literature. *Clin Orthop* 1982; 166:72–74.

130. Keene JS: Low back pain in the athlete from spondylogenic injury during recreation or competition. *Postgrad Med* 1983; 74:209–217.

131. Hadley LA: Stress fracture with spondylolysis. *AJR* 1963; 90:1258–1262.

132. Murray RO, Colwill MR: Stress fracture of the pars interarticularis. *Proc R Soc Med* 1968; 61:555–557.

133. Letts M, Smallman T, Afanasiev R, et al: Fracture of the pars interarticularis in adolescent athletes: A clinical-biomechanical analysis. *J Pediatr Orthop* 1986; 6:40–46.

134. Ferguson RJ, McMaster JH, Stanitski CL: Low back pain in college football linemen. *Am J Sports Med* 1975; 2:63–69.

135. Wertzberger KL, Peterson HA: Acquired spondylo-

lysis and spondylolisthesis in the young child. *Spine* 1980; 5:437–442.

136. Rabushka SE, Apfelbach H, Love L: Spontaneous healing of spondylolysis of the fifth lumbar vertebra: Case report. *Clin Orthop* 1973; 93:256–259.

137. Gelfand MJ, Strife JL, Kereiakes JG: Radionuclide bone imaging in spondylolysis of the lumbar spine in children. *Radiology* 1981; 140:191–195.

138. Elliott S, Hutson MA, Wastie JL: Bone scintigraphy in the assessment of spondylolysis in patients attending a sports injury clinic. *Clin Radiol* 1988; 39:269–272.

139. Wilkinson RH, Hall JE: The sclerotic pedicle: Tumor or pseudotumor. *Radiology* 1974; 111:683–688.

140. Sherman FC, Wilkinson RH, Hall JE: Reactive sclerosis of a pedicle and spondylolysis in the lumbar spine. *J Bone Joint Surg [Am]* 1977; 59:49–54.

141. Horne J, Cockshott WP, Shannon HS: Spinal column damage from water-skiing. *Skeletal Radiol* 1987; 16:612–616.

142. Wassman K: Kyphosis juvenilis Scheuermann: An occupational disorder. *Acta Orthop Scand* 1951; 21:65–74.

143. Kernahan M, Kirpatrick J, Stanish WD: An investigation into the incidence of low back pain in horseback riders. *Nova Scotia Med Bull* 1979; 167–169.

144. Chism SE, Soule AB: Snowmobile injuries. *JAMA* 1969; 209:1672–1674.

145. Dzioba RB: Irreversible spinal deformity in Olympic gymnasts. *Orthop Trans* 1984; 8:66.

146. Keats TE: *Atlas of Normal Roentgen Variants That May Simulate Disease*, ed 4. Chicago, Year Book Medical Publishers, 1988.

147. Meurman KOA: Stress fracture of the pubic arch in military recruits. *Br J Radiol* 1980; 53:521–524.

148. Ozburn MS, Nichols JW: Pubic ramus and adductor insertion stress fractures in female basic trainees. *Milit Med* 1981; 146:332–334.

149. Pavlov H, Nelson TL, Warren RF, et al: Stress fractures of the pubic ramus. *J Bone Joint Surg [Am]* 1972; 64:1021–1025.

150. Noakes TD, Smith JA, Lindenberg G, et al: Pelvic stress fractures in long distance runners. *Am J Sports Med* 1985; 13:120–123.

151. Kim SM, Park CH, Gartland JJ: Stress fracture of the pubic ramus in a swimmer. *Clin Nucl Med* 1987; 12:118–119.

152. Moran JJM: Stress fractures in pregnancy. *Am J Obstet Gynecol* 1988; 158:1274–1277.

153. Wells J: Osteitis condensans ilii. *AJR* 1956; 76:1141–1143.

154. Death B, Kirby RL, MacMillan CL: Pelvic ring mobility: Assessment by stress radiography. *Arch Phys Med Rehab* 1982; 63:204–206.

155. Chamberlain WE: The symphysis pubis in the roentgen examination of the sacroiliac joint. *AJR* 1930; 24:621–625.

156. Clancy WG Jr, Foltz AS: Iliac apophysitis and stress fractures in adolescent runners. *Sports Med* 1976; 4:214–218.

157. Lombardo SJ, Retting AC, Kerlan RK: Radiographic abnormalities of the iliac apophysis in adolescent athletes. *J Bone Joint Surg [Am]* 1983; 65:444–446.

158. Fernbach SK, Wilkinson RH: Avulsion injuries of the pelvis and proximal femur. *AJR* 1981; 137:581–584.

159. Metzmaker JN, Pappas AM: Avulsion fractures of the pelvis. *Am J Sports Med* 1985; 13:349–358.

160. Wojtys EM: Sports injuries in the immature athlete. *Orthop Clin North Am* 1987; 18:689–708.

161. Winkler AR, Barnes JC, Ogden JA: Break dance hip: Chronic avulsion of the anterior superior iliac spine. *Pediatr Radiol* 1987; 17:501–502.

162. Pavlov H, Torg JS: *The Running Athlete: Roentgenograms and Remedies*. Chicago, Year Book Medical Publishers, 1987.

163. Young LW, Tan KM: Traumatic ischial apophyseolysis. *AJDC* 1980; 134:885–886.

164. Kjaerulff H, Hejgaard N, Rostgaard E: Necrosis of the tuberosity of the ischium mimicking neoplasm. *Injury* 1985; 16:554–556.

165. Howse AJG: Osteitis pubis in an olympic roadwalker. *Proc R Soc Med* 1964; 57:88–90.

166. Muckle DS: Associated factors in recurrent groin and hamstring injuries. *Br J Sports Med* 1982; 16:37–39.

167. Cochrane GM: Osteitis pubis in athletes. *Br J Sports Med* 1971; 5:233–235.

168. Hanson PG, Angevine M, Juhl JH: Osteitis pubis in sports activities. *Physician Sports Med* 1978; 6:111–114.

169. Wiley JJ: Traumatic osteitis pubis: The gracilis syndrome. *Am J Sports Med* 1983; 11:360–363.

170. Peirson EL Jr: Osteochondritis of the symphysis pubis. *Surg Gynecol Obstet* 1929; 49:834–838.

171. Schneider R, Kaye JJ, Ghelman B: Adductor avulsive injuries near the symphysis pubis. *Radiology* 1976; 120:567–569.

172. Rold JF, Rold BA: Pubic stress symphysitis in a female distance runner. *Physician Sports Med* 1986; 14:61–65.

173. Tehranzadeh J, Kurth LA, Elyaderani MK, et al: Combined pelvic stress fracture and avulsion of the adductor longus in a middle-distance runner: A case report. *Am J Sports Med* 1982; 10:108–111.

174. Long MM, Stetts DM: Stress fractures of the femoral neck. *Orthop Nurs* 1985; 4:69–71, 76.

175. Orava S: Stress fractures. *Br J Sports Med* 1980; 14:40–44.

176. Ernst J: Stress fracture of the neck of the femur. *J Trauma* 1964; 4:71–83.

177. Fullerton LR Jr, Snowdy HA: Femoral neck stress fractures. *Am J Sports Med* 1988; 16:365–377.

178. Lombardo SJ, Benson DW: Stress fractures of the femur in runners. *Am J Sports Med* 1982; 10:219–227.

179. Case 9. *Radiographics* 1988; 8:253–255.

180. Baer S, Shakespeare D: Stress fracture of the femoral neck in a marathon runner. *Br J Sports Med* 1984; 18:42–43.

181. Hajek MR, Noble HB: Stress fractures of the femoral neck in joggers: Case reports and review of the literature. *Am J Sports Med* 1982; 10:112–116.

182. Saunders AJ, El Sayed TF, Hilson AJ, et al: Stress lesions of the lower leg and foot. *Clin Radiol* 1979; 30:649–651.

183. Miller F, Wenger DR: Femoral neck stress fracture in a hyperactive child: A case report. *J Bone Joint Surg [Am]* 1979; 61:435–437.

184. Wolfgang GL: Stress fracture of the femoral neck in a patient with open capital femoral epiphyses. *J Bone Joint Surg [Am]* 1977; 59:680–681.

185. Gaucher A, Colomb JN, Naoun A, et al: Radionuclide imaging in hip abnormalities. *Clin Nucl Med* 1980; 5:214–226.

186. El Khoury GY, Wehbe MA, Bonfiglio M, et al: Stress fractures of the femoral neck: A scintigraphic sign for early diagnosis. *Skeletal Radiol* 1981; 6:271–273.

187. Coldwell D, Gross GW, Boal DK: Stress fracture of the femoral neck in a child (stress fracture). *Pediatr Radiol* 1984; 14:174–176.

188. Pitt MJ, Graham AR, Shipman JH, et al: Herniation pit of the femoral neck. *AJR* 1982; 138:1115–1121.

189. Barth E, Hagen R: Juxta-articular bone cyst. *Acta Orthop Scand* 1982; 53:215–217.

190. Orava S, Virtanen K: Osteochondroses in athletes. *Br J Sports Med* 1982; 16:161–168.

191. Lindholm TS, Osterman K, Vankka E: Osteochondritis dissecans of elbow, ankle and hip: A comparison survey. *Clin Orthop* 1980; 147:245–253.

192. Butler JE, Brown SL, McConnell BG: Subtrochanteric stress fractures in runners. *Am J Sports Med* 1982; 10:228–238.

193. Luchini MA, Sarokhan AJ, Micheli LJ: Acute displaced femoral-shaft fractures in long-distance runners: Two case reports. *J Bone Joint Surg [Am]* 1983; 65:689–691.

194. Provost RA, Morris JM: Fatigue fracture of the femoral shaft. *J Bone Joint Surg [Am]* 1969; 51:487–498.

195. Blatz DJ: Bilateral femoral and tibial shaft stress fractures in a runner. *Am J Sports Med* 1981; 9:322–325.

196. Levin DC, Blazina ME, Levine E: Fatigue fractures of the shaft of the femur. *Radiology* 1967; 89:883–885.

197. Godshall RW, Hansen CA, Rising DC: Stress fractures through the distal femoral epiphysis in athletes: A previously unreported entity. *Am J Sports Med* 1981; 9:114–116.

198. Ihmeidan IH, Tehranzadeh J, Oldham SA, et al: Case report 443. *Skeletal Radiol* 1987; 16:581–583.

199. Ballmer PE, Bessler WT: Case report 495. *Skeletal Radiol* 1988; 17:382–384.

200. Boade WA: Early confirmation of injury in joggers. *JAMA* 1980; 244:1436.

201. Keats TE, Joyce JM: Metaphyseal cortical irregularities in children: A new perspective on a multifocal growth variant. *Skeletal Radiol* 1984; 12:112–118.

202. Brower AC, Culver JE Jr, Keats TE: Histological nature of the cortical irregularity of the medial posterior distal femoral metaphysis in children. *Radiology* 1971; 99:389–392.

203. Resnick D, Greenway G: Distal femoral cortical defects, irregularities, and excavations: A critical review of the literature with the addition of histologic and paleopathologic data. *Radiology* 1982; 143:345–354.

204. Bufkin WJ: The avulsive cortical irregularity. *AJR* 1971; 112:487–489.

205. Burrows PE, Greenberg ID, Reed MH: The distal femoral defect: Technetium-99m pyrophosphate bone scan results. *J Can Assoc Radiol* 1982; 33:91–93.

206. Barnes GR Jr, Gwinn JL: Distal irregularities of the femur simulating malignancy. *AJR* 1974; 122:180–185.

207. Houston AN, Roy WA, Faust RA, et al: Pellegrini-Stieda syndrome: Report of 44 cases followed from original injury. *South Med J* 1968; 61:113–117.

208. Liu R, Chou C, Yeh S: Three-phase bone scintigraphy in Pellegrini-Stieda disease. *Clin Nucl Med* 1987; 12:47–49.

209. Cahill B: Treatment of juvenile osteochondritis dissecans and osteochondritis dissecans of the knee. *Clin Sports Med* 1985; 4:367–384.

210. Stougaard J: Familial occurrence of osteochondritis dissecans. *J Bone Joint Surg [Br]* 1964; 45:542–543.

211. Phillips HO IV, Grubb SA: Familial multiple osteochondritis dissecans: Report of a kindred. *J Bone Joint Surg [Am]* 1985; 67:155–156.

212. Aichroth P: Osteochondritis dissecans of the knee: A clinical survey. *J Bone Joint Surg [Br]* 1971; 53:440–447.

213. Outerbridge RE: Osteochondritis dissecans of the posterior femoral condyle. *Clin Orthop* 1983; 175:121–129.

214. Mubarak SJ, Carroll NC: Juvenile osteochondritis dissecans of the knee: Etiology. *Clin Orthop* 1981; 157:200–211.

215. Cahill B, Berg G: 99m-Technetium phosphate compound joint scintigraphy in the management of juvenile osteochondritis dissecans of the femoral condyles. *Am J Sports Med* 1983; 11:329–335.

216. Cayea PD, Pavlov H, Sherman MF, et al: Lucent articular lesion in the lateral femoral condyle: Source of patellar femoral pain in the athletic adolescent. *AJR* 1981; 137:1145–1149.

217. Mollan AB: Osteochondritis dissecans of the knee: A case report of an unusual lesion on the lateral femoral condyle. *Acta Orthop Scand* 1977; 48:517–519.

218. Fornasier VL, Czitrom AA, Evans JA, et al: Case report 398. *Skeletal Radiol* 1987; 16:57–59.

219. Hendryson IE: Bursitis in the region of the fibular collateral ligament. *J Bone Joint Surg* 1946; 28:446–450.

220. Liljedahl SO: Common injuries in connection with conditioning exercises. *Scand J Rehab Med* 1971; 3:1–5.

221. Renne JW: The iliotibial band friction syndrome. *J Bone Joint Surg [Am]* 1975; 57:1110–1111.

222. Smillie IS: *Injuries of the Knee Joint*, ed 4. New York, Churchill Livingstone, Inc, 1973.

223. Dickason JM, Fox JM: Fracture of the patella due to overuse syndrome in a child: A case report. *Am J Sports Med* 1982; 10:248–249.

224. Iwaya T, Takatori Y: Lateral longitudinal stress fracture of the patella: Report of three cases. *J Pediatr Orthop* 1985; 5:73–75.

225. Hughes AW: Case report: Avulsion fracture involv-

ing the body of the patella. *Br J Sports Med* 1985; 19:119–120.

226. Kelly DW, Carter VS, Jobe FW, et al: Patellar and quadriceps tendon ruptures — jumper's knee. *Am J Sports Med* 1984; 12:375–380.

227. Vainionpaa S, Bostman O, Patiala H, et al: Megapatella following a rupture of patellar tendon. *Am J Sports Med* 1985; 13:204–205.

228. Kaye JJ, Freiberger RH: Fragmentation of the lower pole of the patella in spastic lower extremities. *Radiology* 1971; 101:97–100.

229. Zernicke RF, Garhammer J, Jobe FW: Human patellar-tendon rupture: A kinetic analysis. *J Bone Joint Surg [Am]* 1977; 59:179–183.

230. Feldman R, Sedlin ED: Ligamentous disruption of the knee associated with avulsion of the patellar tendon in a 12-year-old boy. *J Pediatr Orthop* 1983; 3:101–103.

231. Gould ES, Taylor S, Naidich JB, et al: MR appearance of bilateral, spontaneous patellar tendon rupture in systemic lupus erythematosus. *J Comput Assist Tomogr* 1987; 11:1096–1097.

232. Kricun R, Kricun ME, Arangio GA, et al: Patellar tendon rupture with underlying systemic disease. *AJR* 1980; 135:803–807.

233. Halpern AA, Horowitz BG, Nagel DA: Tendon ruptures associated with corticosteroid therapy. *West J Med* 1977; 127:378–382.

234. Ford LT, DeBender J: Tendon rupture after local steroid injection. *South Med J* 1979; 72:827–830.

235. Haswell DM, Berne AS, Graham CB: The dorsal defect of the patella. *Pediatr Radiol* 1976; 4:238–242.

236. Goergen TG, Resnick D, Greenway G, et al: Dorsal defect of the patella (DDP): A characteristic radiographic lesion. *Radiology* 1979; 130:333–336.

237. van Holsbeeck M, Vandamme B, Marchal G, et al: Dorsal defect of the patella: Concept of its origin and relationship with bipartite and multipartite patella. *Skeletal Radiol* 1987; 16:304–311.

238. Ogden JA, McCarthy SM, Jokl P: The painful bipartite patella. *J Pediatr Orthop* 1982; 2:263–269.

239. Rideout DF, Davis S, Navani SV: Osteochondritis dissecans patellae. *Br J Radiol* 1966; 39:673–675.

240. Edwards DH, Bentley G: Osteochondritis dissecans patellae. *J Bone Joint Surg [Br]* 1977; 59:58–63.

241. Pantazopoulos T, Exarchou E: Osteochondritis dissecans of the patella: Report of four cases. *J Bone Joint Surg [Am]* 1971; 53:1205–1207.

242. McBryde AM Jr: Stress fractures in runners, in D'Ambrosia R, Drez D Jr (eds): *Prevention and Treatment of Running Injuries.* Thorofare, NJ, Charles B Slack, 1982, pp 22–42.

243. Cahill BR: Stress fracture of the proximal tibial epiphysis: A case report. *Am J Sports Med* 1977; 5:186–187.

244. Engber WD: Stress fractures of the medial tibial plateau. *J Bone Joint Surg [Am]* 1977; 59:767–769.

245. Harolds JA: Fatigue fractures of the medial tibial plateau. *South Med J* 1981; 74:578–581.

246. Trimmings NP: An unusual stress fracture of the upper end of the tibia:A case report. *Injury* 1985; 16:348–349.

247. Daffner RH, Martinez S, Gehweiler JA Jr, et al: Stress fractures of the proximal tibia in runners. *Radiology* 1982; 142:63–65.

248. Waisman Y, Varsano I, Grunebaum M, et al: Stress fractures: A diagnostic problem. *Arch Dis Child* 1987; 62:847–848.

249. Collier BD, Johnson RB, Carrera GF, et al: Scintigraphic diagnosis of stress-induced incomplete fractures of the proximal tibia. *J Trauma* 1984; 24:156–160.

250. Orava S, Hulkko A: Stress fracture of the midtibial shaft. *Acta Orthop Scand* 1984; 55:35–37.

251. Blank S: Transverse tibial stress fractures: A special problem. *Am J Sports Med* 1987; 15:597–602.

252. Daffner RH: Anterior tibial striations. *AJR* 1984; 143:651–653.

253. Rettig AC, Shelbourne KD, McCarroll JR, et al: The natural history and treatment of delayed union stress fractures of the anterior cortex of the tibia. *Am J Sports Med* 1988; 16:250–255.

254. Burrows HJ: Fatigue infraction of the middle of the tibia in ballet dancers. *J Bone Joint Surg [Br]* 1956; 38:83–94.

255. Brahms MA, Fumich RM, Ippolito VD: Atypical stress fracture of the tibia in a professional athlete. *Am J Sports Med* 1980; 8:131–132.

256. Devas MB: Stress fractures of the tibia in athletes or "shin soreness." *J Bone Joint Surg [Br]* 1958; 40:227–239.

257. Subbarao K: Radiologic problem of the month: Stress fractures involving anterior tibial cortex. *NY State J Med* 1980; 80:1419–1420.

258. Moore MP: Shin splints: Diagnosis, management, prevention. *Postgrad Med* 1988, 83:199–210.

259. Mills GQ, Marymont JH III, Murphy DA: Bone scan utilization in the differential diagnosis of exercise-induced lower extremity pain. *Clin Orthop* 1980; 149:207–210.

260. Zlatkin MB, Bjorkengren A, Sartoris DJ, et al: Stress fractures of the distal tibia and calcaneus subsequent to acute fractures of the tibia and fibula. *AJR* 1987; 149:329–332.

261. Michael RH, Holder LE: The soleus syndrome: A cause of medial tibial stress (shin splints). *Am J Sports Med* 1985; 15:87–94.

262. Milgrom C, Giladi M, Stein M, et al: Medial tibial pain: A prospective study of its cause among military recruits. *Clin Orthop* 1986; 213:167–171.

263. Mubarak SJ, Gould RN, Lee YF, et al: The medial tibial stress syndrome: A cause of shin splints. *Am J Sports Med* 1982; 10:201–205.

264. Chisin R, Milgrom C, Giladi M, et al: Clinical significance of nonfocal scintigraphic findings in suspected tibial stress fractures. *Clin Orthop* 1987; 220:200–205.

265. Milgrom C, Chisin R, Giladi M, et al: Negative bone scans in impending tibial stress fractures: A report of three cases. *Am J Sports Med* 1984; 12:488–491.

266. Smith JA: Periosteal elevation in a 2 1/2-year-old child. *JAMA* 1976; 236:2323–2324.

267. Walter NE, Wolf MD: Stress fractures in young athletes. *Am J Sports Med* 1977; 5:165–170.

268. Engh CA, Robinson RA, Milgram J: Stress fractures

in children. *J Trauma* 1970; 10:532–541.

269. Davies AM, Evans N, Grimer RJ: Fatigue fractures of the proximal tibia simulating malignancy. *Br J Radiol* 1988; 61:903–908.

270. Dunbar JS, Owen HF, Nogrady MB, et al: Obscure tibial fracture of infants: The toddler's fracture. *J Can Assoc Radiol* 1964; 15:136–144.

271. Ozonoff MB: *Pediatric Orthopedic Radiology.* Philadelphia, WB Saunders Co, 1979.

272. Price AE, Evanski PM, Waugh TR: Bilateral simultaneous Achilles tendon ruptures: A case report and review of the literature. *Clin Orthop* 1986; 213:249–250.

273. Holt JF: Physiological bowing of the legs in young children. *JAMA* 1954; 154:390–394.

274. Bateson EM: The relationship between Blount's disease and bow legs. *Br J Radiol* 1968; 41:107–114.

275. Bateson EM: Non-rachitic bow leg and knock-knee deformities in young Jamaican children. *Br J Radiol* 1966; 39:92–101.

276. Langenskiold A, Riska EB: Tibia vara (osteochondrosis deformans tibiae): A survey of 71 cases. *J Bone Joint Surg [Am]* 1964; 46:1405–1420.

277. Golding JSR, McNeil-Smith JDG: Observations on the etiology of tibia vara. *J Bone Joint Surg [Br]* 1963; 45:320–325.

278. Currarino G, Kirks DR: Lateral widening of epiphyseal plates in knees of children with bowed legs. *AJR* 1977; 129:309–312.

279. Meyers MH, McKeever FM: Fracture of the intercondylar eminence of the tibia. *J Bone Joint Surg [Am]* 1959; 41:209–222.

280. McDonnell MF, Butler JE III: Ostoechondral fracture of the tibial plateau in a ballerina. *Am J Sports Med* 1988; 16:417–418.

281. Hand WL, Hand CR, Dunn AW: Avulsion fractures of the tibial tubercle. *J Bone Joint Surg [Am]* 1971; 53:1579–1583.

282. Ogden JA, Tross RB, Murphy MJ: Fractures of the tibial tuberosity in adolescents. *J Bone Joint Surg [Am]* 1980; 62:205–215.

283. Lipscomb AB, Gilbert PP, Johnston RK, et al: Fracture of the tibial tuberosity with associated ligamentous and meniscal tears. *J Bone Joint Surg [Am]* 1984; 66:790–792.

284. Ogden JA, Southwick WO: Osgood-Schlatter's disease and tibial tuberosity development. *Clin Orthop* 1976; 116:180–189.

285. Scotti DM, Sadhu VK, Heimberg F, et al: Osgood-Schlatter's disease: An emphasis on soft tissue changes in roentgen diagnosis. *Skeletal Radiol* 1979; 4:21–25.

286. Bowers KD: Patellar tendon avulsion as a complication of Osgood-Schlatter's disease. *Am J Sports Med* 1981; 9:356–359.

287. Levine AH, Pais MJ, Berinson H, et al: The soleal line: A cause of tibial pseudoperiostitis. *Radiology* 1976; 119:79–81.

288. Devas MB, Sweetnam R: Stress fractures of the fibula: A review of 50 cases in athletes. *J Bone Joint Surg [Br]* 1956; 38:818–829.

289. Blair WF, Hanley SR: Stress fracture of the proximal fibula. *Am J Sports Med* 1980; 8:212–213.

290. Symeonides PP: High stress fractures of the fibula. *J Bone Joint Surg [Br]* 1980; 67:192–193.

291. Ingersoll CF: Ice skater's fracture: A form of fatigue fracture. *AJR* 1943; 50:469–479.

292. Read MT: Runner's stress fracture produced by an aerobic dance routine. *Br J Sports Med* 1984; 18:40–41.

293. Hopson CN, Perry DR: Stress fractures of the calcaneus in women marine recruits. *Clin Orthop* 1977; 128:159–162.

294. Doury P, Pattin S, Granier R, et al: Donnees nouvelles sur les "fractures de fatigue": A propos d'une observation de fracture de fatigue bilaterale de l'astragale, interet de la scintigraphie osseuse dans le diagnostic des fractures de fatigue. *Rev Rhum Mal Osteoartic* 1984; 51:483–486.

295. Starshak RJ, Simons GW, Sty JR: Occult fracture of the calcaneus: Another toddler's fracture. *Pediatr Radiol* 1984; 14:37–40.

296. Buchanan J, Greer RB III: Stress fractures in the calcaneus of a child: A case report. *Clin Orthop* 1978; 135:119–120.

297. Stein RE, Stelling FH: Stress fracture of the calcaneus in a child with cerebral palsy. *J Bone Joint Surg [Am]* 1977; 59:131.

298. Perry DR, O'Toole ED: Stress fracture of the talar neck and distal calcaneus. *J Am Podiatr Med Assoc* 1981; 71:637–638.

299. Doury P, Pattin S, Granier R, et al: Fractures de fatigue inhabituelles interessant a la fois le calcaneum et le tibia chez une jeune recrue: Interet de la scintigraphie osseuse aux polyphosphates de technetium 99m. *Med Armees* 1981; 9:227–230.

300. Haglund P: Beitrag zur Klinik der Achillesehne. *Z Orthop* 1928; 49:49.

301. Pavlov H, Heneghan MA, Hersh A, et al: The Haglund syndrome: Initial and differential diagnosis. *Radiology* 1982; 144:83–88.

302. Reinig JW, Dorwart RH, Roden WC: MR imaging of a ruptured Achilles tendon. *J Comput Assist Tomogr* 1985; 9:1131–1134.

303. Leekam RN, Salsberg BB, Bogoch E, et al: Sonographic diagnosis of partial Achilles tendon rupture and healing. *J Ultrasound Med* 1986; 5:115–116.

304. Campbell G, Warnekros W: A tarsal stress fracture in a long-distance runner: A case report. *J Am Podiatr Med Assoc* 1983; 73:532–535.

305. Doury P, Pattin S, Metges PJ, et al: A propos d'une observation exceptionnelle de fracture de fatigue bilaterale de l'astragale: Donnees nouvelles sur les fractures de fatigue. *Presse Med* 1984; 13:885–886.

306. Meurman KOA: Stress lesions of the talus. *ROFO* 1980; 132:469–471.

307. Keats TE, Harrison RB: Hypertrophy of the talar beak. *Skeletal Radiol* 1979; 4:37–39.

308. McDougall A: Footballer's ankle. *Lancet* 1955; 269:1219–1220.

309. McMurray JP: Footballer's ankle. *J Bone Joint Surg [Br]* 1950; 32:68–73.

310. Berndt AL, Harty M: Transchondral fractures (osteochondritis dissecans) of the talus. *J Bone Joint Surg [Am]* 1959; 41:988–1020.

311. Newberg AH: Osteochondral fractures of the dome

of the talus. *Br J Radiol* 1979; 52:105–109.

312. DeGinder WL: Osteochondritis dissecans of the talus. *Radiology* 1955; 65:590–598.

313. Mannis CI: Transchondral fracture of the dome of the talus sustained during weight training. *Am J Sports Med* 1983; 11:354–355.

314. Scharling M: Osteochondritis dissecans of the talus. *Acta Orthop Scand* 1978; 49:89–94.

315. Roden S, Tillegard P, Unander-Scharin L: Osteochondritis dissecans and similar lesions of the talus: Report of 55 cases with special reference to etiology and treatment. *Acta Orthop Scand* 1953; 23:51–66.

316. Zinman C, Wolfson N, Reis ND: Osteochondritis dissecans of the dome of the talus: Computed tomography scanning in diagnosis and follow-up. *J Bone Joint Surg [Am]* 1988; 70:1017–1019.

317. Torg JS, Pavlov H, Torg E: Overuse injuries in sport: The foot. *Clin Sports Med* 1987; 6:291–320.

318. Goergen TG, Venn-Watson EA, Rossman DJ, et al: Tarsal navicular stress fractures in runners. *AJR* 1981; 136:201–203.

319. Pavlov H, Torg JS, Freiberger RH: Tarsal navicular stress fractures: Radiographic evaluation. *Radiology* 1983; 148:641–645.

320. Waugh W: The ossification and vascularisation of the tarsal navicular and their relation to Kohler's disease. *J Bone Joint Surg [Br]* 1958; 40:765–777.

321. Nicastro JF, Haupt HA: Probable stress fracture of the cuboid in an infant: A case report. *J Bone Joint Surg [Am]* 1984; 66:1106–1108.

322. Meurman KO: Less common stress fractures in the foot. *Br J Radiol* 1981; 54:1–7.

323. Meurman KO, Elfving S: Stress fracture of the cuneiform bones. *Br J Radiol* 1980; 53:157–160.

324. Maseritz IH: March foot associated with undescribed changes of the internal cuneiform and metatarsal bones. *Arch Surg* 1936; 32:49–64.

325. Childress HM: March fractures of the lower extremity: Report of a case of march fracture of a cuneiform bone. *War Med* 1943; 4:152–160.

326. Levy JM: Stress fractures of the first metatarsal. *AJR* 1978; 130:679–681.

327. Black JR: Stress fractures of the foot in female soldiers: A 2-year survey. *Milit Med* 1982; 147:861–862.

328. Warren MP, Brooks-Gunn J, Hamilton LH, et al: Scoliosis and fractures in young ballet dancers: Relation to delayed menarche and secondary amenorrhea. *N Engl J Med* 1986; 314:1348–1353.

329. Ford LT, Gilula LA: Stress fractures of the middle metatarsals following the Keller operation. *J Bone Joint Surg [Am]* 1977; 59:117–118.

330. Percy EC, Gamble FO: An epiphyseal stress fracture of the foot and shin splints in an anomalous calf muscle in a runner. *Br J Sports Med* 1980; 14:110–113.

331. Hulkko A, Orava S, Nikula P: Stress fracture of the fifth metatarsal in athletes. *Ann Chir Gynaecol* 1985; 74:233–238.

332. Cimmino CV: The Morton's toes. *V Med* 1982; 109:333–334.

333. Frede TE, Lee JK: Compensatory hypertrophy of bone following surgery on the foot. *Radiology* 1983; 146:347–348.

334. Schneider HJ, King AY, Bronson JL, et al: Stress injuries and developmental change of lower extremities in ballet dancers. *Radiology* 1974; 113:627–632.

335. Binek R, Levinsohn EM, Bersani F, et al: Freiberg disease complicating unrelated trauma. *Orthopedics* 1988; 11:753–757.

336. Maurice HD, Newman JH, Watt I: Bone scanning of the foot for unexplained pain. *J Bone Joint Surg [Br]* 1987; 69:448–452.

337. Chillag K, Grana WA: Medial sesamoid stress fracture. *Orthopedics* 1985; 8:819, 821.

338. Helms CA: Radiology of jogger's injuries. *Orthopedics* 1982; 5:1492–1504.

339. Van Hal ME, Keene JS, Lange TA, et al: Stress fractures of the great toe sesamoids. *Am J Sports Med* 1982; 10:122–128.

340. Hulkko A, Orava S, Pellinen P, et al: Stress fractures of the sesamoid bones of the first metatarsophalangeal joint in athletes. *Arch Orthop Traum Surg* 1985; 104:113–117.

341. Feldman F, Pochaczevsky R, Hecht H: The case of the wandering sesamoid and other sesamoid afflictions. *Radiology* 1970; 96:275–283.

342. Pessina R: Os peroneum fracture: A case report. *Clin Orthop* 1988; 277:261–264.

343. Golding C: The sesamoid of hallux: Museum pages. *J Bone Joint Surg [Br]* 1960; 42:840–843.

344. Tehranzadeh J, Stoll DA, Gabriele OM: Case report 271. *Skeletal Radiol* 1984; 12:44–47.

345. Scranton PE Jr, Rutkowski R: Anatomic variations in the first ray: II. Disorders of the sesamoids. *Clin Orthop* 1980; 151:256–264.

346. Ogata K, Sugioka Y, Urano Y, et al: Idiopathic osteonecrosis of the first metatarsal sesamoid. *Skeletal Radiol* 1986; 15:141–145.

347. Pentecost RL, Murray RA, Brindley HH: Fatigue, insufficiency, and pathologic fractures. *JAMA* 1964; 187:1001–1004.

348. Demartini F, Grokoest AW, Ragan C: Pathological fractures in patients with rheumatoid arthritis treated with cortisone. *JAMA* 1952; 149:750–752.

349. Gershon-Cohen J, Rechtman AM, Schraer H, et al: Asymptomatic fractures in osteoporotic spines of the aged. *JAMA* 1953; 153:625–627.

350. Kaplan PA, Orton DF, Assleson RJ: Osteoporosis with vertebral compression fractures, retropulsed fragments, and neurologic compromise. *Radiology* 1987; 165:533–535.

351. Simmons CR, Harle TS, Singleton EB: The osseous manifestations of leukemia in children. *Radiol Clin North Am* 1968; 6:115–130.

352. Samuda GM, Cheng MY, Yeung CY: Back pain and vertebral compression: An uncommon presentation of childhood acute lymphoblastic leukemia. *J Pediatr Orthop* 1987; 7:175–178.

353. Resnick D, Niwayama G: *Diagnosis of Bone and Joint Disorders*, ed 2. Philadelphia, WB Saunders Co, 1988, p 3059.

354. Cooper KL: Insufficiency fractures of the sternum: A consequence of thoracic kyphosis? *Radiology* 1988; 167:471–472.

355. Baum ML, Kramer EL, Sanger JJ, et al: Stress frac-

tures and reduced bone mineral density with prior anorexia nervosa, letter. *J Nucl Med* 1987; 28:1506–1507.

356. Rigotti NA, Nussbaum SR, Herzog DB, et al: Osteoporosis in women with anorexia nervosa. *N Engl J Med* 1984; 311:1601–1606.

357. Rigotti NA, Near RM, Jameson L: Osteopenia and bone fractures in a man with anorexia nervosa and hypogonadism. *JAMA* 1986; 256:385–388.

358. Brotman AW, Stern TA: Osteoporosis and pathologic fractures in anorexia nervosa. *Am J Psychiatry* 1985; 142:495–496.

359. Davies AM, Evans NS, Struthers GR: Parasymphyseal and associated insufficiency fractures of the pelvis and sacrum. *Br J Radiol* 1988; 61:103–108.

360. Lourie H: Spontaneous osteoporotic fracture of the sacrum: An unrecognized syndrome of the elderly. *JAMA* 1982; 248:715–717.

361. Ries T: Detection of osteoporotic sacral fractures with radionuclides. *Radiology* 1983; 146:783–785.

362. Schneider R, Yacovone J, Ghelman B: Unsuspected sacral fractures: Detection by radionuclide bone scanning. *AJR* 1985; 144:337–341.

363. De Smet AA, Neff JR: Pubic and sacral insufficiency fractures: Clinical course and radiologic findings. *AJR* 1985; 145:601–606.

364. Taylor RT, Huskisson EC, Whitehouse GH, et al: Spontaneous fractures of pelvis in rheumatoid arthritis. *Br Med J* 1971; 4:663–664.

365. Rafii M, Firooznia H, Golimbu C, et al: Bilateral acetabular stress fractures in a paraplegic patient. *Arch Phys Med Rehabil* 1982; 63:240–241.

366. Guha SC, Poole MD: Stress fracture of the iliac bone with subfascial femoral neuropathy: Unusual complications at a bone graft donor site. *Br J Plast Surg* 1983; 36:305–306.

367. Casey D, Mirra J, Staple TW: Parasymphyseal insufficiency fractures of the os pubis. *AJR* 1984; 142:581–586.

368. Godfrey N, Staple TW, Halter D, et al: Insufficiency os pubis fractures in rheumatoid arthritis. *J Rheumatol* 1985; 12:1176–1179.

369. Goergen TG, Resnick D, Riley RR: Post-traumatic abnormalities of the pubic bone simulating malignancy. *Radiology* 1978; 126:85–87.

370. Hall FM: Postfracture pubic osteolysis simulating malignancy. *AJR* 1984; 143:433.

371. Resnick D, Guerra J Jr: Stress fractures of the inferior pubic ramus following hip surgery. *Radiology* 1980; 137:335–338.

372. Launder WJ, Hungerford DS: Stress fracture of the pubis after total hip arthroplasty. *Clin Orthop* 1981; 159:183–185.

373. Oh I, Hardacre JA: Fatigue fracture of the inferior pubic ramus following total hip replacement for congenital hip dislocation. *Clin Orthop* 1980; 147:154–156.

374. Resnick D, Guerra J Jr: Stress fractures associated with adjacent osteoarthritis. *J Rheumatol* 1981; 8:161–164.

375. Torisu T: Fatigue fracture of the pelvis after total knee replacements: Report of a case. *Clin Orthop* 1980; 149:216–219.

376. Cracchiolo A: Stress fractures of the pelvis as a cause of hip pain following total hip and knee arthroplasty. *Arthritis Rheum* 1981; 24:740–742.

377. Gaucher A, Pourel J, Widerkehr P, et al: Fractures de contrainte apres osteotomie tibiale. *Rev Rhum Mal Osteoartic* 1981; 48:253–256.

378. Hooyman JR, Melton LJ III, Nelson AM, et al: Fractures after rheumatoid arthritis: A population-based study. *Arthritis Rheum* 1984; 27:1353–1361.

379. Miller B, Markheim HR, Towbin MN: Multiple stress fractures in rheumatoid arthritis: A case report. *J Bone Joint Surg [Am]* 1967; 49:1408–1414.

380. Pullar T, Parker J, Capell HA: Spontaneous fractured neck of femur in rheumatoid arthritis: Absence of radiographic changes on initial X-ray. *Scott Med J* 1985; 30:178–180.

381. Scott JA, Rosenthal DI, Chandler HP: Stress fracture of the femoral neck following internal fixation: A case report. *Injury* 1986; 17:419–420.

382. McElwaine JP, Sheehan JM: Spontaneous fractures of the femoral neck after total replacement of the knee. *J Bone Joint Surg [Br]* 1982; 64:323–325.

383. Lesniewski PJ, Testa NN: Stress fracture of the hip as a complication of total knee replacement: Case report. *J Bone Joint Surg [Am]* 1982; 64:304–306.

384. Lotke PA, Wong RY, Ecker ML: Stress fracture as a cause of chronic pain following revision total hip arthroplasty. *Clin Orthop* 1986; 206:147–150.

385. Perry HM III, Weinstein RS, Teitelbaum SL, et al: Pseudofractures in the absence of osteomalacia. *Skeletal Radiol* 1982; 8:17–19.

386. Orwoll ES, McClung MR: Pseudofractures in patients with low-turnover osteoporosis. *West J Med* 1985; 143:239–242.

387. Richardson RMA, Rapoport A, Oreopoulos DG, et al: Unusual fractures associated with osteoporosis in premenopausal women. *Can Med Assoc J* 1978; 119:473–476.

388. Manson D, Martin RF, Cockshott WP: Metaphyseal impaction fractures in acute lymphoblastic leukemia. *Skeletal Radiol* 1989; 17:561–564.

389. Scott RD, Turoff N, Ewald FC: Stress fracture of the patella following duopatellar total knee arthroplasty with patellar resurfacing. *Clin Orthop* 1982; 170:147–151.

390. Bauer G, Gustafsson M, Mortensson W, et al: Insufficiency fractures in the tibial condyles in elderly individuals. *Acta Radiol [Diagn]* 1981; 22:619–622.

391. Manco LG, Schneider R, Pavlov H: Insufficiency fractures of the tibial plateau. *AJR* 1983; 140:1211–1215.

392. Reynolds MT: Stress fractures of the tibia in the elderly associated with knee deformity. *Proc R Soc Med* 1972; 65:377–380.

393. Satku K, Kumar VP, Pho RW: Stress fractures of the tibia in osteoarthritis of the knee. *J Bone Joint Surg [Br]* 1987; 69:309–311.

394. Martin LM, Bourne RB, Rorabeck CH: Stress fractures associated with osteoarthritis of the knee. *J Bone Joint Surg [Am]* 1988; 70:771–774.

395. Rand JA, Coventry MB: Stress fractures after total knee arthroplasty. *J Bone Joint Surg [Am]* 1980; 62:226–233.

396. Neuwirth M: Stress fracture after high tibial osteotomy: A case report. *Bull Hosp J Dis Orthop Inst* 1978; 39:92–99.

397. Haider R, Storey G: Spontaneous fractures in rheumatoid arthritis. *Br Med J* 1962; 1:1514–1516.

398. Fam AG, Shuckett R, McGillivray DC, et al: Stress fractures in rheumatoid arthritis. *J Rheumatol* 1983; 10:722–726.

399. Schneider R, Kaye JJ: Insufficiency and stress fractures of the long bones occurring in patients with rheumatoid arthritis. *Radiology* 1975; 116:595–599.

400. Young A, Kinsella P, Boland P: Stress fractures of the lower limb in patients with rheumatoid arthritis. *J Bone Joint Surg [Br]* 1981; 63:239–243.

401. Ross DJ, Dieppe PA, Watt I, et al: Tibial stress fracture in pyrophosphate arthropathy. *J Bone Joint Surg [Br]* 1983; 65:474–477.

402. Schnitzler CM, Solomon L: Trabecular stress fractures during fluoride therapy for osteoporosis. *Skeletal Radiol* 1985; 14:276–279.

403. Miniaci A, McLaren AC, Haddad RG: Longitudinal stress fracture of the tibia: Case report. *J Can Assoc Radiol* 1988; 39:221–223.

404. Schnitzler CM, Solomon L: Histomorphometric analysis of a calcaneal stress fracture: A possible complication of fluoride therapy for osteoporosis. *Bone* 1986; 7:193–198.

405. Leroux JL, Blotman F, Claustre J, et al: Fractures du calcaneum au cours du traitement de l'osteoporose par le fluor. *Sem Hop Paris* 1983; 59:3140–3142.

406. Baron M, Paltiel H, Lander P: Aseptic necrosis of the talus and calcaneal insufficiency fractures in a patient with pancreatitis, subcutaneous fat necrosis, and arthritis. *Arthritis Rheum* 1984; 27:1309–1313.

407. Lauzon C, Carette S, Mathon G: Multiple tendon rupture at unusual sites in rheumatoid arthritis. *J Rheumatol* 1987; 14:369–371.

408. Milkman LA: Multiple spontaneous idiopathic symmetrical fractures. *AJR* 1934; 32:622–634.

409. Steinbach HL, Kolb FO, Gilfillan R: A mechanism of the production of pseudofractures in osteomalacia (Milkman's syndrome). *Radiology* 1954; 62:388–395.

410. Steinbach HL, Noetzli M: Roentgen appearance of the skeleton in osteomalacia and rickets. *AJR* 1964; 91:955–972.

411. Griffin CN Jr: Symmetrical ilial pseudofractures: A complication of chronic renal failure. *Skeletal Radiol* 1982; 8:295–298.

412. Tarr RW, Kaye JJ, Nance EP Jr: Insufficiency fractures of the femoral neck in association with chronic renal failure. *South Med J* 1988; 81:863–866.

413. Skinner HB, Harris JR, Cook SD, et al: Bilateral sequential tibial and fibular fatigue fractures associated with aluminum intoxication osteomalacia: A case report. *J Bone Joint Surg [Am]* 1983; 65:843–847.

414. Barry HC: Orthopedic aspects of Paget's disease of bone. *Arthritis Rheum* 1980; 23:1128–1130.

415. Stevens J: Orthopaedic aspects of Paget's disease. *Metab Bone Dis Relat Res* 1981; 4-5:271–278.

416. Milgram JW: Radiographical and pathological assessment of the activity of Paget's disease of bone. *Clin Orthop* 1977; 127:43–54.

417. Looser E: Osteitis deformans und unfall. *Arch Klin Chir* 1934; 180:379–386.

418. Dove J: Complete fractures of the femur in Paget's disease. *J Bone Joint Surg [Br]* 1980; 62:12–17.

419. Matin P, Lang G, Carretta R, et al: Scintigraphic evaluation of muscle damage following extreme exercise: Concise communication. *J Nucl Med* 1983; 24:308–311.

420. Barloon TJ, Zachar CK, Harkens KL, et al: Rhabdomyolysis: Computed tomography findings. *CT: J Comput Tomogr* 1988; 12:193–195.

421. Sagar VV, Meckelnburg RL, Chaikin HL: Bone scan in rhabdomyolysis. *Clin Nucl Med* 1980; 5:321–322.

422. Kaplan GN: Ultrasonic appearance of rhabdomyolysis. *AJR* 1980; 134:375–377.

423. Hellmann DB, Helms CA, Genant HK: Chronic repetitive trauma: A cause of atypical degenerative joint disease. *Skeletal Radiol* 1983; 10:236–242.

424. Mintz G, Fraga A: Severe osteoarthritis of the elbow in foundry workers. *Arch Environ Health* 1973; 27:78–80.

425. Williams WV, Cope R, Gaunt WD, et al: Metacarpophalangeal arthropathy associated with manual labor (Missouri metacarpal syndrome): Clinical, radiographic, and pathologic characteristics of an unusual degenerative process. *Arthritis Rheum* 1987; 30:1362–1371.

426. Rifkin MD, Levine RB: Driller wrist (vibratory arthropathy). *Skeletal Radiol* 1985; 13:59–61.

427. Burke MJ, Fear EC, Wright V: Bone and joint changes in pneumatic drillers. *Ann Rheum Dis* 1977; 36:276–279.

428. Kumlin T, Wiikeri M, Sumari P: Radiological changes in carpal and metacarpal bones and phalanges caused by chain saw vibration. *Br J Ind Med* 1973; 30:71–73.

429. Brodelius A: Osteoarthrosis of the talar joints in footballers and ballet dancers. *Acta Orthop Scand* 1960; 31:309–314.

430. Coste F, Desoille R, Illouz G, et al: Appareil locomoteur et danse classique. *Rev Rhum Mal Osteoartic* 1960; 27:259.

431. Odekerken JC, Chantraine A, Bernard A: Osteoarthritis and axis deviation of the knee joint in old athletes (French). *J Belge Rhum Med Phys* 1973: 28:74–85.

432. Thomas RB: Chronic injury: Lacrosse, in Larson LA (ed): *Encyclopedia of Sport Sciences and Medicine*. New York, Macmillan Publishing Co, 1971, p 621.

433. Solonen KA: The joints of the lower extremities of football players. *Ann Chir Gynaecol* 1966; 55:176–180.

434. Rall KL, McElroy GL, Keats TE: A study of the long-term effects of football injury to the knee. *Mo Med* 1964; 61:435–438.

435. Bagneres H: Lesions osteo-articulares chroniques des sportifs. *Rheumatologie* 1967; 19:27–34.

436. Louyot P, Savin R: La coxarthrose chez l'agriculteur. *Rev Rhum Mal Osteoartic* 1966; 33:625.

437. Lawrence JS: Rheumatism in coal miners: III. Occupational factors. *Br J Ind Med* 1955; 12:249.

438. Iselin M: Importance de l'arthrose dans le syndrome main fragile des boxeurs. *Rev Rhum Mal Osteoartic* 1960; 27:242.

439. Layani F, Roeser J, Nadaud M: Les lesions osteo-articulaires des catcheurs. *Rev Rhum Mal Osteoartic* 1960, 27:244.

440. Lawrence JS: Rheumatism in cotton operatives. *Br J Ind Med* 1961; 18:270–276.

441. Vere-Hodge N: Chronic injury: Cricket, in Larson AL (ed): *Encyclopedia of Sport Sciences and Medicine.* New York, Macmillan Publishing Co, 1971, p 605.

442. Bard CC, Sylvester JJ, Dussault RG: Hand osteoarthropathy in pianists. *J Can Assoc Radiol* 1984; 35:154–158.

443. Elliott GB, Elliott KA: The torture or stretch arthritis syndrome (a modern counterpart of the medieval "manacles" and "rack"). *Clin Radiol* 1979; 30:313–315.

Index